Code of Practice
for Electromagnetic
Resilience

Published by The Institution of Engineering and Technology, London, United Kingdom

The Institution of Engineering and Technology is registered as a Charity in England & Wales (no. 211014) and Scotland (no. SC038698).

© The Institution of Engineering and Technology 2017

First published 2017

The Institution of Engineering and Technology,
Michael Faraday House,
Six Hills Way, Stevenage,
SG1 2AY, United Kingdom.

Copies of this publication may be obtained from:
The Institution of Engineering and Technology
PO Box 96, Stevenage, SG1 2SD, UK
Tel: +44 (0)1438 767328
Email: sales@theiet.org
www.electrical.theiet.org/books

ISBN 978-1-78516-324-1 (paperback)
ISBN 978-1-78516-325-8 (electronic)

CONTENTS

ACKNOWLEDGEMENTS

Participants in the Technical Committee
Working Group Members:

Keith Armstrong (**Technical Author**)	Cherry Clough Consultants Ltd
Rhodri Morgan	DNV GL Group, also representing 'The 61508 Association'
Pete Dorey	TUV-Sud
Steve Hayes	EMC Test Labs Association
Dr. Richard Hoad	QinetiQ Group plc
Terry Dunford	Civil Aviation Authority (CAA)
Adam Bullivant	Ricardo Rail
Dr. Brian Kirk	Robinsons Associates
John Cryer	Office of the Nuclear Regulator (ONR)

Other contributors and correspondents

Geoff South	BAE Systems
Prof. Chris Jones	BAE Systems
Prof. Ian MacDiarmid	Ex BAE Systems
Will Turner	MPE Ltd
Ken Webb	Mott McDonald Rail
Dr. Bill Radasky	Metatech Corporation, USA
Prof. Davy Pissoort	University of Leuven (KU Leuven), Belgium
Prof. Alistair Duffy	IEEE/De Montford University
Ron Bell, OBE	Ex HSE, Ron Bell Consulting Ltd
Alan Warner	EMC Industries Association
Doug Nix	Compliance InSight Consulting Inc.
Stephen Zerenner	Siemens Healthcare Diagnostics Inc.

The IET also wishes to thank Barry Lytollis for his original work on this subject.

SECTION 1

Introduction

1.1 Overview

This Code of Practice provides guidance on the assessment and application of techniques and measures that can address the interfering effects of electromagnetic disturbances on safety-related systems.

When competently selected and applied, a set of such techniques and measures will provide part of the evidence required for justifying functional safety decisions and for compliance with functional safety standards (including IEC 61508 Ed.2:2010, or functional safety standards that are based on IEC 61508).

The scope of this Code of Practice does not cover human health issues caused by electromagnetic fields (EMF).

1.2 Why manage electromagnetic resilience?

All electronic and electro-mechanical technologies can suffer errors, malfunctions or failures due to electromagnetic interference (EMI), which can be caused by electromagnetic disturbances. For non-safety equipment, functionality in this regard is generally demonstrated by compliance with all applicable electromagnetic immunity test standards.

Where electronic or electro-mechanical technologies are used in a safety-related application, it is necessary to ensure that EMI cannot cause safety risks to exceed tolerable limits. However, compliance with electromagnetic immunity test standards (for example, those listed as providing a presumption of conformity to the EMC directive) does not demonstrate safety performance, and traditional immunity testing cannot validate equipment's safety performance to the confidence required for safety applications.

In practice a number of factors related to electromagnetic disturbances limit the confidence with which safety performance can be demonstrated, including:

(a) over a lifecycle, in any real situation the characteristics and locations of sources of electromagnetic disturbances can rarely be known with confidence, and often cannot be predicted with complete confidence.

(b) operational experience is of very limited value in demonstrating safety performance with respect to EMI throughout the operation of a safety-related system. Errors, malfunctions and failures caused by EMI are often transitory, leaving no tangible evidence of their occurrence, making fault identification and evidence gathering extremely difficult unless specific techniques and measures are used.

(c) no practicable amount of electromagnetic immunity testing can be sufficient, on its own, to demonstrate the performance of a safety component or system throughout its life.

The likelihood of EMI impacting upon the performance of safety-related systems is also increasing because:

(a) there is a rapidly increasing use of wireless systems, high-speed and/or high-power switching devices and programmable electronics, in all applications. The electromagnetic disturbances increasingly being emitted by these technologies, and their increasing use, will make the electromagnetic environment progressively worse, in terms of noise levels, spectral density and bandwidth for the foreseeable future.

(b) modern safety systems increasingly employ technologies with the potential for decreased immunity (increased susceptibility), for example, modern electronics operating at lower switching voltages, lower power consumption, and wireless receivers. The application of such technologies may not be evident without detailed analysis of product design information.

(c) the use of intentionally-generated electromagnetic disturbances to prevent the operation of a system is a growing concern, not just for the military sector but also for critical infrastructure and other sectors.

Consequently, there is a need to manage the electromagnetic resilience of safety systems, in order to ensure that they are capable of delivering their intended risk reduction throughout their operation.

The basic IEC publication covering electromagnetic compatibility (EMC) for functional safety is IEC 61000-1-2:2016 [2]. This publication establishes a methodology through which the functional safety of electromagnetic phenomena can be achieved for electrical and electronic systems and installations.

IEC 61000-1-2:2016 [2] uses the terminology of IEC 61508 and can be used to support claims, made in accordance with IEC 61508 and functional safety standards based upon it, that functional safety has been achieved with regard to electromagnetic disturbances.

The medical industry uses IEC 60601-1-2 Ed.4:2014 to manage the risks that can be caused by electromagnetic disturbances.

Electromagnetic resilience

The methodology set out in IEC 61000-1-2:2016 [2] is based on risk management principles and focuses on reducing risks throughout the lifecycle of a safety-related system. Table 3 and Annex B within that Standard briefly describe a number of techniques and measures related to electromagnetic resilience — the subject of this Code of Practice — as being a practical method for achieving compliance. Figure 1.1 provides an overview of this from a management perspective.

Functional Safety

This is the part of the overall safety of a system that depends upon the correct functioning of electrical and/or electronic equipment

The Basic Publication on Functional Safety is IEC 61508

Electromagnetic Interference (EMI)

Some natural phenomena (lightning, static discharges, etc.) and *all* electrical and electronic equipment emit certain EM disturbances

All electrical and electronic equipment (hardware, software, systems, installations, etc.) is vulnerable to certain EM disturbances

When they are so affected, we say they are suffering from EMI

Functional Safety Risks due to EMI

Where EMI may cause errors, malfunctions or failures in the correct functioning of electrical or electronic equipment (hardware and/or software), this can in turn increase functional safety risks

Note: this is not concerned with compliance with the EMC Directive or any other EMC regulations; only with functional safety risks

The IEC's Basic Publication on achieving Functional Safety despite EM disturbances is IEC 61000-1-2:2016

Reduction of Risks due to EMI

Requires the use of appropriate design, verification and validation techniques and measures from concept to completion

Also in maintenance, repair, refurbishment, upgrades, modifications, etc., throughout the lifecycle

Electromagnetic Resilience

Costs and Liabilities

Functional Safety risks caused by EMI can be very costly to deal with, and may also cause breaches of Health & Safety and/or Product Liability legislation

Note: "electrical" equipment includes electromechanical; and "electronic" equipment includes programmable electronic

1.3 The aims and application of this Code of Practice

The aim of this Code of Practice is to support the achievement of functional safety performance by providing guidance on the selection and application of electromagnetic resilience techniques and measures.

Its purpose is to support the adoption of adequate electromagnetic resilience engineering practices throughout the functional safety lifecycle, by offering further guidance and practical advice on the application of risk management activities, including the techniques and measures set out in IEC 61000-1-2:2016 [2].

This Code of Practice is intended to be used by those who have responsibilities for functional safety. While it is aimed at functional safety, the methodologies, techniques and measures described here can also be used for the reduction of other risks, such as security risks and non-safety-related risks (for example, risks to the operation of commercial IT systems).

1.4 Relationship with IEC 61508:2010 and other safety standards

This Code of Practice supplements the information on dealing with EMI given in IEC 61508 and other standards or specifications containing electromagnetic requirements for functional safety, such as IEC 61000-1-2 [2], IEC 61326-3-1 [3], IEC 61326-3-2 [4], IEC 61000-6-7 [5] and IEC 60601-1-2. It also supersedes the information in the IET's 2008 and 2013 guides [6].

Note: IEC 61326-3-1 and IEC 61326-3-2 are not considered complete 'EMC for functional safety' standards because they are not based on IEC 61000-1-2, which is the IEC's basic standard on this topic].

The situation regarding electromagnetic disturbances and functional safety was not well-understood when the basic safety standard IEC 61508 was first published in 2000, or when the functional safety standards based upon IEC 61508 were first published (see Annex C).

The result is that, generally, the EMC requirements in those standards are currently based upon a restricted set of standard laboratory tests, applied only to representative examples of equipment, which cannot provide sufficient confidence that safety-related systems, or the sub-systems or elements they incorporate, will not suffer intolerable risks because of EMI throughout their lifecycle.

IEC 61000-1-2 Ed.1:2016 [2] was created specifically to provide the electromagnetic requirements that are missing from such standards, and the 2nd Edition of IEC 61508 listed an early version, IEC TS 61000-1-2:2008, as a normative reference.

IEC 61000-1-2:2016 [2] is intended by its authoring committee to complement IEC 61508:2010. It provides guidance on how electromagnetic engineering may contribute to the achievement of functional safety, because this area is not covered in detail in IEC 61508:2010. IEC 61000-1-2:2016 is a basic safety standard and, like IEC 61508, is intended to be used by the authors of other functional safety standards as the technical basis for the requirements in the standards they create.

IEC 61000-1-2:2016 describes the situation regarding electromagnetic disturbances and the achievement of functional safety, and its Annex B describes electromagnetic resilience. It includes practical techniques and measures for ensuring that electromagnetic disturbances do not cause functional safety risks to exceed tolerable levels, based upon the IET's 2013 guidance [6].

The relationship of this Code of Practice to IEC and ISO standards has been described above. Where IEC and ISO standards have been adapted to fulfil national or Trade Bloc regulatory purposes (for example, when prefixed with EN, ANSI, etc., or completely renumbered) this Code of Practice can be considered to be relevant to them in the same way.

1.5 Achieving electromagnetic resilience for functional safety

The methodology set out in IEC 61000-1-2 [2] (and, for medical devices, in IEC 60601-1-2 Ed.4:2014) is based on risk management principles and focuses on reducing risks due to electromagnetic disturbances throughout the entire lifecycle, from concept to end of life. The approach taken by this Code of Practice is illustrated in Figure 1.2.

▼ **Figure 1.2** Achieving electromagnetic resilience

Good EMC and functional safety engineering practices used throughout the design, including appropriate techniques and measures

Compliance with EMC test standards for emissions and immunity applicable to the normal EM environments expected to be experienced - over the lifecycle (assuming no faults)

Appropriate additional practices, techniques & measures are used to ensure risks remain tolerable - despite reasonably foreseeable EMI and/or faults over the lifecycle

Result: EM Resilience

The safety integrity achieved is sufficiently resilient to all reasonably foreseeable EM disturbances and faults over the lifecycle

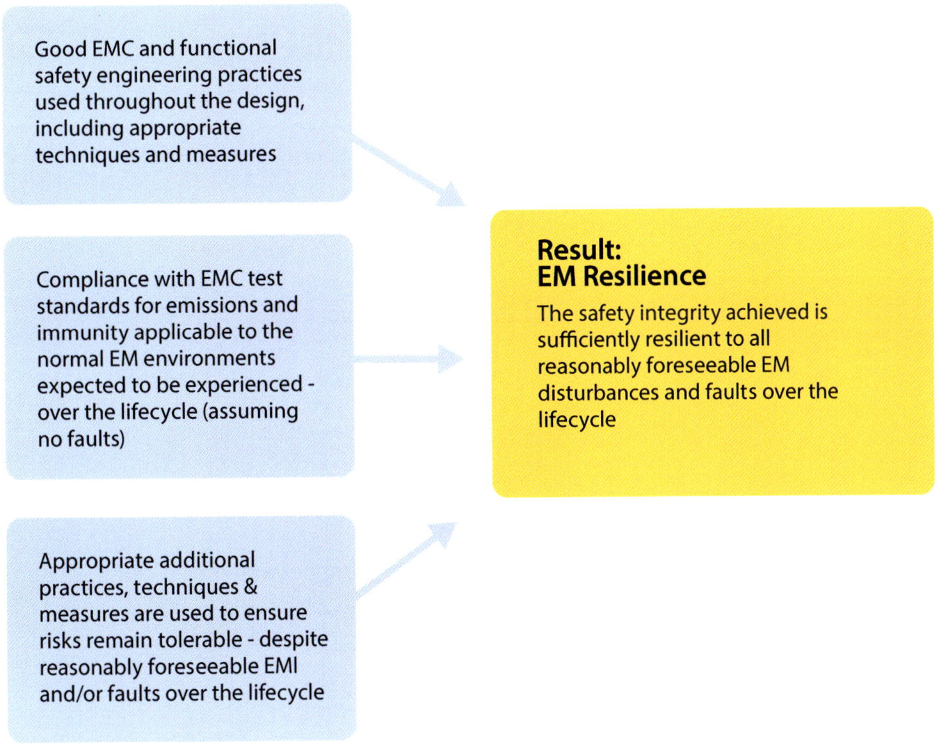

This approach builds on the existing competency of the EMC and functional safety engineering communities, as follows:

(a) Compliance with EMC standards for non-safety-related functionality

EMC design/testing competency for non-safety-related functionality (for example, to comply with the EMC Directive [7], other EMC regulations or customer-specific EMC specifications such as those used by the military) is very well established in the developed world.

The EMC emissions and immunity standards that are used have been developed over the years (and are still developing) for the purpose of ensuring adequate availability of equipment/system functionality, with different standards being developed to suit the wide variety of generalised electromagnetic environments (such as domestic, commercial, light industrial, heavy industrial, military, road vehicles, railways, etc.).

This Code of Practice builds on the above experience, competency, and existing EMC test facilities, by recommending that equipment or systems:

(i) comply with all published test standards relevant for the electromagnetic disturbances that are expected both to occur in the normal electromagnetic environments and to be experienced by the safety-related system, sub-system or element.

(ii) continue to comply with all published relevant test standards throughout their anticipated lifecycles.

Note: These tests may exceed what is required for EMC Directive compliance, in the number of types of electromagnetic disturbance tested, their levels, and their characteristics (for example, for continuous disturbances: modulation types; frequency ranges; dwell times, etc., and for transient disturbances: wave shapes; number of occurrences, etc.).

Note: If the electromagnetic environment experienced by the safety-related system, sub-system or element changes over time, these tests should be reviewed to help ensure that the availability of the equipment under control (EUC) remains adequate.

Functional safety requirements must be met over the complete lifecycle, taking into account all reasonably foreseeable:

(i) operating conditions;

(ii) errors, malfunctions and faults, whether static or intermittent;

(iii) environmental conditions (shock, vibration, humidity, condensation, temperature, electromagnetic disturbances, etc.);

(iv) wear, corrosion, aging, degradations and failures;

(v) component tolerances and variability, construction and installation errors, etc.;

(vi) use and misuse, whether intentional or not;

(vii) multiple electromagnetic disturbances of the same or different types, occurring at the same time or in some critical sequence; and

(viii) any combinations of (i) - (vii) above.

(b) The use of electromagnetic resilience techniques and measures

The level of functional safety competency that is required in order to comply with IEC 61508 and its related standards is well-established worldwide. Generally, the techniques and measures employed by these standards are well understood, well-proven and widely used.

These techniques and measures have been developed to prevent the introduction of systematic errors in the safety-related systems.

Electromagnetic disturbances can also be a cause of errors, malfunctions or faults in any of the electrical or electronic parts of a safety-related system, and such failures are called electromagnetic interference (EMI). Some of the techniques and measures that have been widely used to achieve compliance with IEC 61508 and its related standards may also be partially effective at preventing electromagnetic disturbances from causing EMI that could increase functional safety risks.

This Code of Practice builds on the existing experience and competency in the functional safety engineering community by recommending the use of:

(i) well-established functional safety techniques and measures that are known to be especially effective against the effects of EMI;

(ii) modifications to certain techniques and measures to make them more effective against the effects of EMI;

(iii) the application of good electromagnetic engineering techniques and measures at every level of design (such as components, circuit, PCB, wiring, software, product, element, sub-system and system design, installation design, design for maintainability and repair, etc.) during the anticipated lifecycle; and

(iv) the application of good electromagnetic engineering practices, including appropriate techniques and measures at every stage of the system lifecycle.

IEC 61508 manages the functional safety over the entire lifecycle, and IEC 61000-1-2:2016 gives additional requirements and guidance on functional safety activities relating to electromagnetic disturbances.

This Code of Practice recommends compliance with IEC 61000-1-2, including the application of an adequate range of electromagnetic resilience techniques and measures, in order to achieve an appropriate level of electromagnetic resilience. This Code of Practice provides expanded guidance on electromagnetic resilience techniques and measures, complementary with that published in Table 3 and Annex B of IEC 61000-1-2:2016.

The selection of techniques and measures employed for a particular system will depend on:

(a) the technology employed;

(b) the application; and

(c) the safety integrity level (SIL) of the system as specified in IEC 61508, or their equivalents as specified in a comparable functional safety standard.

Sufficient assessment shall be carried out to justify the selection of electromagnetic resilience techniques and measures for a safety-related system (see Section 1.7).

No single technique or measure can be relied upon alone. The functional safety designer must choose a set that ensures that — regardless of the electromagnetic disturbances that can cause the errors, malfunctions or failures — the overall functional safety specifications are met.

As the required level of risk reduction for a safety-related system increases, so does the safety integrity level from, for example, SIL 1 to SIL 4. Demonstrations of adequacy for higher SIL levels will typically involve more developed technical arguments and documentation than those for lower SIL levels. The range of techniques and measures required, as well as the rigour expected from their demonstration, will depend on the situation, but will usually increase significantly with each SIL level. A similar approach applies to elements, as the systematic capability increases from SC1 to SC4. The competent application of a set of electromagnetic resilience techniques and measures (see Section 2) shall be recorded as part of an overall approach to safety documentation based on IEC 61508 principles. The safety documentation should demonstrate that there is sufficient confidence that functional safety would not be compromised by electromagnetic disturbances over the anticipated lifecycle. The adequacy of the overall safety argument expressed in the safety documentation could be subject to independent assessment, where appropriate.

For equipment intended to be used in a safety-related system (for example, sub-systems or elements), the competent application of electromagnetic resilience techniques and measures should be recorded in the product's documentation, and made available to system integrators, installers, end users, and assessors as necessary. For example, information related to electromagnetic resilience could be included in the safety manual for compliant items required by IEC 61508.

1.6 Application of electromagnetic resilience techniques and measures

This Code of Practice describes a range of techniques and measures that are considered to be useful, where relevant, for improving the electromagnetic resilience during all stages in the safety-related system or safety-related product lifecycle (see Figure 1.3).

IEC 61508 requires all functional safety-related projects to consider all these stages, whether they concern the creation of equipment for use in constructing a safety-related system (usually standard products manufactured in volume) or the integration or creation of a safety-related system itself.

▼ **Figure 1.3**   Overview of the functional safety lifecycle (source: IEC 61508 Part 1 Ed.2:2010)

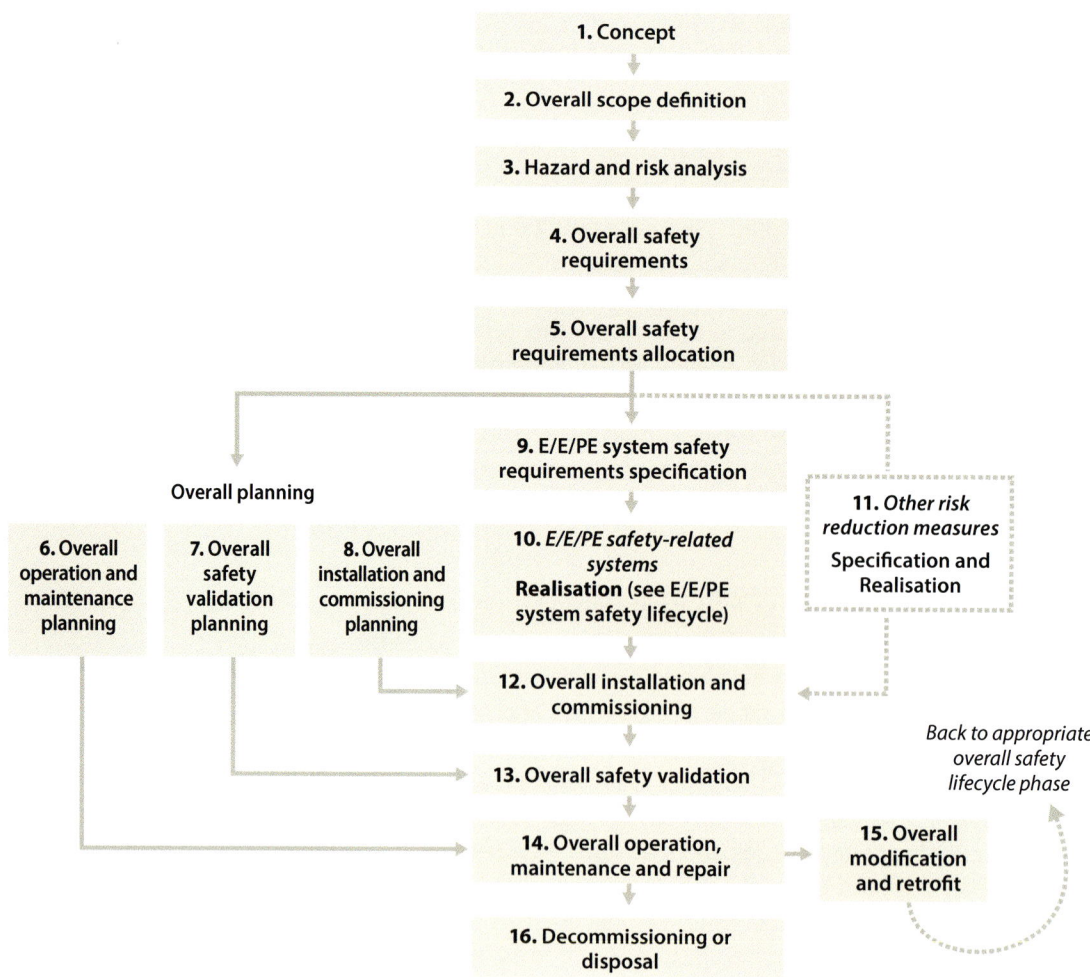

Activities shall be identified and employed to deliver sufficient electromagnetic resilience for functional safety over the anticipated lifecycle of a safety-related system. These activities shall meet the requirements of relevant functional safety standards and IEC 61000-1-2:2016, and should be applied during the following project stages, as necessary:

(a) project management, planning and specification;
(b) system design;
(c) operational design;
(d) system implementation, integration, installation and commissioning;
(e) verification and validation, including through-life monitoring;
(f) operation, maintenance, repair, refurbishment, upgrade as necessary; and
(g) decommissioning as necessary.

These activities should be identified, managed and documented in line with the standards expected for other aspects of functional safety, and should therefore be allocated to competent persons as necessary.

> ▶ Responsibilities for the achievement of functional safety should therefore be identified. This should include the identification of responsibilities for ensuring adequate electromagnetic resilience of the safety-related system or product.

Annex A is a checklist of the electromagnetic resilience techniques and measures that are described in detail in Part 2 of this Code of Practice. This checklist includes columns detailing their importance versus their safety integrity level or safety capability.

1.7 Assessment of electromagnetic resilience techniques and measures

As part of functional safety lifecycle management activities, appropriate arrangements shall be made to select and implement a suitable range of techniques and measures for electromagnetic resilience.

An adequate combination of techniques and measures shall be selected that, together, achieve the required safety integrity level/systematic capability with respect to electromagnetic disturbances. Their selection shall be recorded within the safety documentation. It is recommended that an assessment of the necessary techniques and measures should be prepared. Adequate reasons should be recorded for the selections made and for rejecting those that are not used. Annex A in this Code of Practice provides a basic checklist that may be applied for this purpose.

This Code of Practice identifies a number of techniques and measures that may be employed at appropriate stages of the system lifecycle as necessary. However, those applied need not be limited to those given in this Code of Practice — additional techniques and measures may give added assurance of electromagnetic resilience.

No electromagnetic resilience techniques and measures, such as those described in this Code of Practice, should be assumed to guarantee complete protection against every possible type of electromagnetic disturbance, combination of electromagnetic disturbances, faults or misuse that could result in EMI.

The exact combination of techniques and measures selected for a particular application will depend on many factors specific to the application in question. Except where stated otherwise, the techniques and measures covered by this Code of Practice are appropriate for both 'continuous' and 'on-demand' safety functions.

Depending on the nature of the project, different electromagnetic resilience techniques and measures might be used in its various stages:

(a) if a project did not involve any software design, then no software design techniques and measures would be selected for any of the project's stages; likewise
(b) if there was no circuit design required, then circuit design techniques and measures are not needed.

The extent to which robust conventional electromagnetic compatibility (EMC) management techniques (such as high-specification electromagnetic mitigation including shielding, filtering and transient suppression) can prevent electromagnetic disturbances from affecting the correct operation of a safety-related system during its anticipated lifecycle may be taken into account during the selection and application of the techniques and measures, where this is justified.

Each technique or measure described in this Code of Practice is presented based on its relevance to the stage of the project, under the headings: Aim; Description; Identification; Mitigation, and Importance.

Aim	The overall purpose of the technique or measure.
Description	Broadly how the technique or measure achieves its aim.
Identification	The effectiveness of the technique or measure to reveal the presence of an error or malfunction that could be caused by electromagnetic disturbances.
Mitigation	The behaviour of the system safety function in response to the detected errors or malfunctions that could have been caused by electromagnetic disturbances.

Importance will specify the necessity and/or desirability of the technique or measure for reducing the risks due to electromagnetic disturbances using the attributes: Not Recommended (NR); Recommended (R); Highly Recommended (HR) and Mandatory (M). In this Code of Practice, the 'Importance' of a technique or measure as NR, R, HR or M (see above) is graded according to the relevant safety integrity level (SIL) as specified in IEC 61508, or an equivalent as specified in a related functional safety or risk management standard.

Parts 1 and 5 of IEC 61508 describes its methodology for determining the safety integrity level of a safety function, and other functional safety or risk management standards will have equivalent methodologies. Also see Annex D in Part 7 of IEC 61508.

In accordance with the methodology used in IEC 61508, if a technique or measure rated as HR for the relevant safety integrity level/systematic capability is not used, a detailed technical explanation of why not should be included in the relevant safety documentation. For example, the technique or measure might not actually be relevant for the design being implemented, or it might be that an alternative technique or measure is used instead, which provides the same benefits for risk-reduction to the design issue concerned.

Notes to consider when reading this Code of Practice

Note: For clarity, the Importance for each technique and measure is only shown in the checklist table in Annex A.

Note: Where a technique or measure in Sections 2.2 or 2.3 applies to a technology that is not relevant to the equipment or system concerned, and the importance as shown in Annex A as being M or HR, a justification for why that technique or measure was not applied should be included in the safety documentation (see Section 1.8).

Note: The 'Importance' levels (R, HR, M) listed in Annex A are generic starting points, and an informed application consistent with expectations of the relevant sector should be made. For example, in certain industries (e.g. rail, military, nuclear) where Annex A lists importance as R they might expect HR or M; and where Annex A lists HR they might expect M.

1.8 Documentation of electromagnetic resilience

1.8.1 Documentation of electromagnetic resilience for safety-related systems

In the case of a complete safety-related system, the safety documentation should contain all the necessary evidence that shows that the overall safety-related system is adequate for its required safety duty.

Prior to the safety-related system being put into use, a structured justification of adequate electromagnetic resilience of the system shall be produced. This justification could be assessed as part of functional safety assessment activities; alternatively, other appropriate structured assessment approaches could be used.

This justification should assess the extent of compliance with the requirements of relevant functional safety standards including IEC 61000-1-2:2016, and the adequacy of the range of activities, techniques and measures employed for electromagnetic resilience, including verification and validation activities.

Supporting information should be available for this justification as necessary and this will usually include electromagnetic environment specifications, electromagnetic tests results and certifications, and material demonstrating the range and suitability of the electromagnetic resilience techniques and measures used in the safety lifecycle.

Information used to support claims made as to the safety performance of a safety-related system should be made available to system integrators and end users and should be retained in order to support system review activities throughout the lifecycle of the system.

Independent assessment of the electromagnetic resilience justification should be undertaken where appropriate. The extent of independence required for assessment should be in line with the approach suggested for independent assessment in IEC 61508-1:2010.

1.8.2 Documentation of electromagnetic resilience for equipment (such as system elements)

In the case of equipment intended for use in a safety-related system, where electromagnetic resilience claims are made for the equipment, information supporting these claims shall be made available in the compliant item's safety manual.

This should include information that will be required by the safety-related system integrator, for example, the safety-related environmental tests that the equipment complies with, the functional safety techniques and measures that have been applied in the design of the equipment, guidance on installation, maintenance etc., as necessary.

A safety manual should also include information on equipment behaviour, where appropriate, in the case of failure, for example, 'defined states' (DSs) that the equipment can assume in response to errors or failures due to intolerable EMI, as well as guidance on the application of the equipment.

Information used to support claims made as to the safety performance of a safety-related system element should be made available to system integrators and end users, and should be retained in order to support system review activities throughout the lifecycle of the element's application.

SECTION 2

Detailed guidance on electromagnetic resilience techniques and measures

2.1 Electromagnetic resilience in project management, planning and specification

2.1.1 Techniques and measures for project management and planning

Aim: To avoid failures in the management, planning, selection, design, implementation, commissioning, verification, and maintenance of measures for avoiding and controlling dangerous failures due to electromagnetic disturbances and EMI.

This applies to a whole safety-related system, and to separate parts of a safety-related system.

Description: The processes for the management, planning, selection, design, implementation, commissioning, modification, verification, and maintenance of each safety function should explicitly include electromagnetic resilience measures and should be documented.

A competent person should have the overall responsibility for managing the electromagnetic resilience of the system. Appropriate competency should be made available at all lifecycle stages.

Identification: By independent assessment of the design for conformance with this Code of Practice, see Clause 8 of [8] for guidance on the appropriate level of independence.

Mitigation: By employing the techniques and measures described in this Code of Practice (or equivalent techniques and measures justified in the safety documentation).

2.1.2 Techniques and measures for use when creating a design requirements specification

Aim:

To ensure that the design specification includes requirements for EMI, and that all reasonably foreseeable electromagnetic disturbances and their effects are taken into account in the specification of the system and its sub-systems and elements.

Appropriate techniques and measures shall be defined and used to ensure that the safety-related system shall achieve the required SIL, and all of the sub-systems and elements incorporated within it shall achieve their required systematic capabilities, despite any electromagnetic disturbances over the lifecycle.

Amongst other issues, the following shall be taken into account:

(a) non-operation, when operation is required;
(b) operation, when no operation is required; and
(c) unintended or inaccurate operations.

The specification for electromagnetic resilience techniques and measures shall be (as far as is possible): complete; free from errors and contradictions; and easy to verify.

Description:

The requirements and design specifications shall be defined using a variety of semi-formal and formal modelling techniques, for example those listed in Annex B.15, including a preliminary hazards analysis as a semi-quantitative technique to be used in the initial design process, and Taguchi's 'Design of Experiments' (see [754]) approach to help get a robust design and also to help test for robustness by quickly determining the worst cases where there are multiple orthogonal effects acting.

Whichever techniques are chosen, the potential effects of EMI on the hardware and software shall be taken into account. Typically this might include consideration of the possibility of corruption of data and program memory content, corruption of data in transit on internal or external serial or parallel buses and their consequent effects on the safe operation of the system.

Put more simply: EMI (including intentional EMI (IEMI)) must be considered as contributing towards the risk of a hazard, and its effects either eliminated, mitigated, or accommodated using appropriate techniques and measures, for example, as described in this Code of Practice.

This activity should take fully into account the fact that electromagnetic disturbances and EMI can cause an effectively infinite variety of:

(a) any/all kinds of noisy, degraded, distorted, false, delayed, re-prioritised, overvoltage, etc. controls/signals/data, both intermittently and continuously;
any/all kinds of under/over voltages, noises, dropouts and interruptions, lasting from less than one microsecond to many seconds, minutes, even permanent, in one or any number of AC or DC power supplies, both intermittently and continuously;
any/all kinds of waveform distortions, frequency perturbations in any number of AC power supplies, plus phase and voltage imbalances in multi-phase supplies;
(b) one or more combinations of any of the above, occurring in any number of signal paths or power supplies, simultaneously or in any critical time relationship.

The design requirements specification should state the selection of electromagnetic resilience techniques and measures to be used for achieving adequate electromagnetic resilience for the intended system, sub-system or element to comply with its safety integrity level/systematic capability in its expected operational environment over its lifecycle.

References:

See the list in Annex B.15.

2.1.3 Specifying EMC test standards to help ensure the availability of the EUC

Aims:
To help ensure adequate availability of the equipment under control (EUC), and of its safety-related systems, throughout its lifecycle, so that safety-related systems continue to provide safe operation, taking into account availability, throughput rate, production rate, or other financial or mission-critical requirements.

Description:
(a) To help ensure that both intentional and unintentional electromagnetic emissions, over the lifecycle, do not exceed levels that are likely to affect other equipment.

(b) To help ensure that the reasonably foreseeable normal operational electromagnetic environment does not cause sufficient EMI to activate any safe failure modes, ensuring adequate availability of the EUC over the lifecycle.

Emissions and immunity tests are selected from the IEC (including CISPR) series of EMC emissions and immunity test methods considered appropriate for both the intended application and the expected electromagnetic environment(s) over the lifecycle (see Annex B.5). The assessment of the expected electromagnetic environment should include both inter-system and intra-system electromagnetic energy coupling paths.

However, other types of EMC tests might be more appropriate than IEC or CISPR, especially for automotive, rail, aerospace, military, etc., applications and environments for which specific EMC test standards have been developed (see Annex B.6-B.12).

To correspond to the predicted electromagnetic environment and the application, the test standards may need to be modified, for example an emissions limit might need to be reduced over a certain frequency range because of the close proximity of certain sensitive equipment, or might need to be extended by some GHz to help protect certain wireless communications.

Example 1

A safety-related system in an industrial plant located near to an airport or harbour might apply IEC 61000-6-4 and IEC 61000-6-2 (the generic standards for emissions and immunity for the heavy industrial environment). It might also need to be tested for immunity to the various radars it will be exposed to by applying tests using the IEC 61000-4-3 method modified to simulate the nearby radar levels, frequencies, modulations, pulse repetition rates, etc.

Example 2

Most safety-related systems will be exposed to close-proximity transmitting portable electronic devices (T-PEDs), radio-frequency identification (RFID) readers, and/or machine-to-machine (M2M) transmitters, and wireless-data-enabled laptops, tablets, PDAs, e-book readers and the like. Consequently, their immunity should be tested accordingly, probably requiring the application of test standards such as [224] and/or [317], in addition to the other EMC immunity tests that have been selected.

Example 3

Proximity to high-power electrical installations might expose safety-related systems to large magnetic fields, high-amplitude conducted noise at frequencies from DC to at least 10 kHz, and/or high energy radiated and conducted transients requiring appropriate testing in addition to the other EMC immunity tests that have been selected.

Also, an immunity level might need to be increased over a certain frequency range, or extended to higher frequencies, because of the close proximity of certain 'noisy' equipment (for example, radio-frequency materials processing equipment operating with high RF power in an ISM band, or a radio-communications transmitter).

It may also be useful to modify standard testing to ensure that specific aspects of the equipment's performance are adequately tested, for example, by extending test frequencies to ensure that the performance of the system clock is adequately tested. The tables of recommended tests in IEC 61000-1-2 [2] and IEC 61000-6-7 [5] may assist with identifying far-field immunity tests. Near-field immunity testing (for an example, see [317]) may also be appropriate for situations where portable radio transmitters (such as mobile phones, cellphones, WiFi, Bluetooth, etc.) could be in very close proximity to the equipment.

Immunity levels may also need to be increased to account for test measurement uncertainty. Testing at the specified limit only provides a 50 % confidence interval that the immunity level has been applied as there is equal probability that the immunity level applied is plus or minus the required limit. Safety standards such as IEC 61508 do not mandate a particular confidence level for electromagnetic measures but for EMC tests a minimum 95 % confidence interval is recommended. Further guidance is available in the UKAS document LAB 34 *The expression of measurement uncertainty in EMC testing* [10].

Where a customer's contractual EMC test is equivalent to a selected test, or exceeds its requirements, it should replace that selected test.

See Section 1 and [11]-[14] for discussions on the fact that no practicable immunity testing plan can, on its own, demonstrate sufficient confidence that electromagnetic disturbances will not cause unacceptable degradation of functional safety over the lifecycle.

Other appropriate techniques and measures, for example, those described in this Code of Practice, are also needed to achieve functional safety as regards electromagnetic disturbances.

Identification:

A test plan shall be devised by persons competent in applying the selected EMC tests, and verification and validation testing carried out according to this plan.

Verification tests (see Section 2.5) should be applied to all relevant elements of the safety-related system, ideally by their manufacturers, during the integration phase.

Validation tests (see Section 2.5) should preferably be applied to the complete safety-related system, functioning in its final configuration in its intended application and environment.

Where this is not practicable, the standard tests should be applied at the highest practicable level of assembly of the safety-related system or sub-systems and the likely limitations and consequences of the partial testing documented.

In addition, in-situ EMC testing should be carried out where practicable, for example, by using the methodology described in [600].

The immunity tests should show that the system elements or the safety-related system itself are unaffected at the applied test levels (i.e. their good electromagnetic design, plus filtering, shielding, etc. offers adequate protection against the electromagnetic disturbances).

The point of complying with immunity test standards is to maintain the required availability of the EUC and its safety-related systems.

To that end, element functions intended for use in system-safety functions should not fail during these tests — unless they fail to a DS and this situation is adequately addressed in the safety documentation.

For the same reason, safety functions themselves should not be triggered during these tests — unless this is adequately addressed in the safety documentation.

Safety functions should never be inhibited from operating as a result of these tests, which may require the use of certain techniques and measures such as those described in this Code of Practice.

The results of the testing according to the plan should be documented and assessed against the relevant design requirements specification. Unexpected or anomalous behaviour should be investigated, the underlying causes corrected, and the work involved documented.

The tests should be carried out in a manner that provides sufficient confidence that compliance with them will be maintained over the complete lifecycle.

> **Example**
>
> Some manufacturers take equipment that complies with its specified EMC emissions and immunity test standards, artificially age it using well-established acceleration techniques, then retest the aged units to check that they still comply with those EMC test standards.

Mitigation: By competently modifying the design using good electromagnetic engineering practices (see Section 2.3.26) until the test requirements are met in a way that indicates their maintenance over the lifecycle.

Note 1: Compliance with EMC Regulations applicable in the country of application is generally a starting point for this specification exercise, but is almost never sufficient, because complying with the conventional test standards alone is insufficient for electromagnetic resilience (see Section 1 and [11]-[14]).

Note 2: Manufacturers are not necessarily precluded from doing these tests themselves, or constrained to using certain types of third-party test laboratories.

The degree of accuracy, confidence, test accreditation and independence required for these tests is — like most functional safety issues — generally dependent on the safety integrity level/systematic capability.

References: Annex B.14 includes some references on assessing electromagnetic environments, and some relevant standards from different industries and application areas.

2.1.4 Protecting against high impact, unusual and malicious EMI

Aim: To help achieve functional safety where the occurrence of high impact, unusual and malicious electromagnetic disturbances could reasonably be foreseen and cause temporary disturbance and/or permanent damage to hardware (electronic components, interconnections, etc.).

Description: Examples of unusual EMI includes: very near proximity lightning stroke, unusual electrostatic discharge (ESD) events and transients, such as corona due to a nearby power fault or HV switching event.

Examples of malicious EMI include HEMP, IEMI and jamming of wireless channels (see [661] [38] [659] [660] and [662]).

Implementation: By specifying appropriate environments, selecting appropriate electromagnetic mitigation and resilience techniques and measures and performing appropriate tests, using (for example) the relevant documents and standards listed in Annex B.

A test plan shall be devised by persons competent in applying the selected EMC tests, and verification and validation testing carried out according to this plan. Verification tests (see Section 2.5) shall be applied to all elements of the safety-related system, ideally by their manufacturers, during the integration phase. Validation tests (see Section 2.5) should preferably be applied to the complete safety-related system, in its final configuration in its intended application, running a typical application program. Where this is not practicable the standard tests should be applied at the highest practicable level of assembly of the safety-related system or sub-systems and the likely limitations and consequences of the partial testing documented. Some of these tests may require in-situ testing with electromagnetic disturbances.

Mitigation: Where it is considered necessary to cope with the occurrence of one or more such high-impact electromagnetic disturbances over the lifecycle, appropriate mitigation should be applied, for example, as described in documents listed in Annexes B.2 or B.3, to pass the relevant tests.

Alternatively, appropriate techniques and measures could be applied to detect inhibition or false operation of the safety function and cause it to default to a redundant or backup safety-related system. To aid fault attribution and diagnostics the fault detection should be correlated to independent EMI event detection and monitoring (see section 2.2.9).

The redundant or backup safety-related system could be normally completely disengaged from all power and signals, so that it is more likely to survive these powerful electromagnetic events and minimise common-cause failures.

A redundant or backup safety-related system that uses low-technology electronics (i.e. does not use programmable electronics) is more likely to survive such powerful electromagnetic events, with a non-electrical backup system likely to be the most rugged.

A 'non-electrical backup system' is one based on mechanical, hydraulic and/or pneumatic technologies alone (i.e. with no electrical or electronic control).

Note: The military and defence sectors have their own sets of standards for these high-power electromagnetic disturbances. See the relevant references in Annex B for examples.

2.2 Electromagnetic resilience techniques and measures for use in system design

This section describes various techniques and measures to help prevent electromagnetic disturbances from degrading the safety integrity of the safety-related system.

During the operation of a system, EMI may cause hardware malfunction in the form of corruption of data in memories, and corruption of signals on data, address and control bus lines and interfaces. This in turn can cause software, and hence the system, to malfunction, possibly presenting a system safety hazard. Techniques and measures should be applied accordingly, bearing in mind all the possible susceptibilities of the system to the variety of electromagnetic disturbances described in Section 2.1.

Some suitable techniques and measures are described in Sections 2.3-2.8, or alternatives may be used if technical justifications are provided in the safety documentation.

Note: Where a technique or measure in this section applies to a technology that is not relevant to the equipment or system concerned, and the importance as shown in Annex A as being M or HR, a justification for why that technique or measure was not applied should be included in the safety documentation (see Section 1.8).

2.2.1 Separating safety-related system parts from non-safety-related parts

Aim: To separate the safety-related parts of a system from non-safety-related parts, such that the electromagnetic disturbances created by the non-safety-related parts, or the consequences of EMI occurring in the non-safety-related parts, do not affect the safety-related parts.

Description: In the specification, it should be decided whether a complete or partial separation of the safety-related systems and non-safety-related systems is possible.

Clear specifications should be written for the interfacing of the two parts.

Possible remaining routes for interference that could create coupling between the safety-related part and the non-safety-related parts should be identified and documented, such that appropriate techniques and measures can be implemented to address them.

Reference: [125]

Note: This technique concerns the physical separation of hardware and the connections made between hardware elements (i.e. their communication, power and physical interfaces).

2.2.2 Recording how the design requirements are implemented through design choices

Aim: To produce a stable design of the safety-related system, and any part of it, in conformance with its design specification (see Section 2.1).

Description: This is where the design choices, mitigation strategies, techniques and their justifications for the electromagnetic resilience techniques, and the measures used to comply with the design specification, are documented.

These will typically include EMI filtering, separation, segregation, grounding and shielding, sufficient at least to meet normal requirements for electromagnetic immunity, together with a selection of techniques and measures such as those described in this Code of Practice according to their importance for the required safety integrity level/systematic capability. See also Section 2.3.26.

The safety documentation shall include a list of all the applicable techniques and measures. This should record the justification for not implementing any rated HR importance (see Section 1.7). The safety documentation should show that the electromagnetic resilience requirements described in the design requirements specification relating to the required safety integrity level/systematic capability are fulfilled.

The checklists in Annex A provide a non-exhaustive selection of techniques and measures that are likely to be applicable during the design process and for the practical implementation. Additional techniques may be used if justified in the safety documentation.

Note 1: It is generally impractical to demonstrate/verify/validate that a set of electromagnetic mitigation techniques and measures alone is sufficient for any particular safety integrity level/systematic capability.

Note 2:   The degree of competence, amount of detail, amount of work, and amount of documentation involved in the above shall be commensurate with the safety integrity level/systematic capability.

2.2.3 Co-design electromagnetically diverse hardware/ software in redundant channels

Aim: To detect and/or correct systematic failures using multiple electromagnetically diverse hardware channels and/or software components, to reduce the likelihood that the common-cause characteristics of electromagnetic disturbances will cause an incorrect output to be created.

Electromagnetically diverse hardware and software designs have different modes and rates of failure due to electromagnetic disturbances.

IEC 61508 describes hardware and software diversity as being different types of techniques and measures. However, these days some traditional hardware diversity techniques and measures may be more effectively accomplished in software, and some traditional software diversity techniques and measures may now be more effectively accomplished in hardware (for example, by using field-programmable gate arrays (FPGAs)) — so co-design is required.

Hardware and software designers should work together (i.e. co-design) to achieve the required overall diversity in the most effective way in order to meet the requirements of the design requirements specification and its required safety integrity levels and/or systematic capabilities.

Diverse hardware: Where a safety-related system uses redundant hardware 'channels' with comparison or voting on their outputs to detect and/or correct errors or faults, these channels should be electromagnetically diverse.

This reduces the probability of systematic common cause errors or failures when the safety-related system experiences electromagnetic disturbances, and increases the probability of detecting such errors and failures, surviving them and maintaining availability.

Methods for achieving electromagnetically diverse hardware channels include (but are not limited to):

(a)   Different physical principles, such as sensing different but related physical parameters, for example, the temperature and pressure of a sealed vessel; using resistances and thermocouple voltages to measure temperature; etc.

(b)   Different digital architectures, such as using processors with different internal structures.

(c)   Algorithms that use different techniques to solve the same problem or calculate the same results.

(d)   Different methods of physical realisation, such as using shielded cables, wireless or fibre-optics for communications.

(e)   Spatial separation, so that an electromagnetic disturbance or ionizing radiation track is likely to only affect one of the redundant channels.

(f)   Locating each item of equipment in a different electromagnetic environment.

(g)   Routing cables such that each cable runs through a different electromagnetic environment.

(h)   Different circuit design principles, such as operating on a signal, the value of which is represented as either a voltage; current; frequency; mark-space ratio; digital code, etc.

(i)   Functional diversity, i.e. the use of different approaches to achieve the same result, such as analogue, digital or optical electronic technologies.

(j) Mechanical, hydraulic and pneumatic technologies have the advantage of being immune to all EMI and may be used to great benefit in some situations.

(k) Inversion of data or signals.

(l) Where different channels are synchronised to the same clock, operating them out of step with each other. Ideally, operating redundant channels non-synchronously.

(m) Where different communication channels, sensors, etc., use specific narrowband frequencies, ensure that each of them uses frequencies that are not harmonically related to the others. Examples include linear variable displacement transducers (LVDTs), strain gauges and other bridge measurements run on AC, Doppler sensors for velocity, metal detectors, solid-state gyroscopes, and any sensor, transducer or other type of circuit that uses phase-sensitive detection, phase-locked loops, or very narrow band-pass filters.

(n) Provide different channels with power from different, independent sources.

An example of using diversity in a multi-channel control system:

Two redundant, identical electronic sensors are mounted on the same printed circuit board, or in the same integrated circuit (IC), and sense the same physical parameter (for example, the position, velocity, temperature, gas concentration, etc.). A comparator checks whether their outputs agree, and switches the EUC into a safe state when they do not.

Because the sensors are so close together, they share the same electromagnetic environment, which means that they experience the same electromagnetic disturbances at the same time.

A common effect of electromagnetic disturbances on electronic sensors is to cause a positive or negative 'zero shift'; when this occurs both of these sensors will give false high or low measurements at the same time.

If large enough, electromagnetic disturbances can cause zero-shifts in many types of sensors of as much as full scale deflection (FSD), but the comparator will be unable to detect any false high or low measurements, even up to ±FSD, because both sensors have the same (false) output at the same time. The EUC would not be switched into a safe state, even if the errors in the sensor signals resulted in unsafe operation (if, for example, the position was too far or not far enough; the velocity, temperature or gas concentration was too high or too low; etc.).

However, introducing electromagnetic diversity by connecting one of the sensors so that it produces signals that are inverted with respect to the other, and restoring the correct polarity at its input to the comparator, makes it highly probable that the sensors' zero-shifts, due to the electromagnetic disturbances, would in fact be detected by the comparator.

There are many other ways of introducing electromagnetic diversity to this simple example.

Diverse software:

The first option for electromagnetic diversity of software is to use two or more independent software components to implement the same safety function, where each component is designed and coded separately and uses different partitions of memory for its data (and may use different algorithms where this is feasible).

Differences in the outputs of these components are detected by the software itself or by means of comparison or voting logic as for hardware redundancy.

The rationale for the use of electromagnetically diverse software components is that a memory corruption or incorrect instruction execution caused by EMI may not affect both (all) of the diverse software components. If it does, then the effects of the EMI will, in general, be different, allowing the comparison or voting logic to detect the error.

The second option for electromagnetically diverse software is to use an electromagnetically diverse monitor: a software component that checks the expected output of the main software against the actual output, to ensure safe (but not necessarily correct) behaviour.

The electromagnetically diverse monitor continually checks the output of the main software and prevents the system entering an unsafe state, either by means of a separate output or by bringing the main software back to a correct state.

An electromagnetically diverse monitor should be simpler than achieving electromagnetically diverse main software. If not, it is equivalent to a redundant implementation.

It may be helpful to implement the electromagnetically diverse monitor on a separate computer to reduce the likelihood of the main software and the diverse software monitor being affected in the same way by the same EMI event.

If a separate computer is not used then the electromagnetically diverse monitor must be capable of operating (and, in particular, capable of recovering from EMI-induced errors) independently of the main software, for example, in a different process or task using separate memory areas.

Electromagnetically diverse software of both kinds may be combined with electromagnetically diverse hardware (using different input channels and/or processors) to further reduce the likelihood of common cause errors due to EMI.

Extending the method to three or more channels requires a voting function that is sufficiently reliable and adequately electromagnetically resilient at the required level of safety. This voter must have a reliability (despite EMI) corresponding to the improvement in confidence that is the purpose of using the multiple channels. Various techniques may be used to do this, for example, dynamic self-testing as described in Section 2.3.21.

Where such voting is used it can be assumed, given sufficient confidence in the electromagnetically diverse behaviour of the channels, that channels that meet the requirements of the voting function are operating correctly. Whilst the voting result is positive the system can maintain the correct operation of the EUC without any need to fail to a safe state.

In the absence of a safe state, the use of a sufficient number of redundant electromagnetically diverse technology channels with a voting function is one of the most important methods for maintaining safety integrity.

Note 1:

Bear in mind that functionally equivalent items of hardware from the same or alternative suppliers may not behave sufficiently differently when subjected to the same electromagnetic disturbances. Their internal hardware and/or software design may not be sufficiently electromagnetically diverse.

Note 2:	It may be possible to suspend the operation of the safety function for a period of time until the channels agree once more, without degrading the safety integrity.
	This helps to maintain availability by reducing the number of times the system fails to a safe state as the result of temporary or transient EMI, and so reduces the possibility that users will modify the system to compromise the correct operation of the safety function (an example of foreseeable misuse).
Note 3:	EMI may cause software instructions or data to change, due to corruption of instruction address and/or data bus.
References:	Methods of partitioning software on the same computer:
	[113]-[116], Annex F of [118], [122]-[127]
	Common cause failures: [746]

2.2.4 System integration, installation and commissioning

Aim:	To ensure that electromagnetic resilience is correctly considered when parts of the system that have been separately tested are brought together to form the complete functional system.
Description:	Most systems are constructed from a variety of functional modules and multiple commercially sourced products.
	Each part needs to be designed and verified as being resilient to EMI, however, further attention is needed when the individual parts of the system are housed and connected, including the shared power supplies and system interconnections that may create additional opportunities for EMI to occur or its effects to be propagated within the system.
	Typical system-level EMI issues might occur through, for example, the inappropriate selection of cable types; cable segregation issues (such as crosstalk); unsuitable earthing/grounding structures; common cause failures due to EMI, etc.
	The approach taken to avoid an increase in system-wide EMI vulnerability due to system integration (physical, electrical and functional) should be documented in the safety documentation.
Identification:	By independent assessment of the design and realisation of the integration against relevant good electromagnetic engineering practices for systems and installations (see Section 2.3.26).
	Clause 8 of IEC 61508 [8], especially its Tables 4 and 5, provide guidance on the independence required for the assessment according to the safety integrity level/systematic capability.
	The use of event data recorders within the system may help to pinpoint the likely causes of malfunction, (see Section 2.2.5), and data communication error counts may provide an indication of EMI influencing communications networks or systems.
Mitigation:	By modification of the relevant design.

2.2.5 Fault detection and event data recording for later diagnosis

Aim: To increase the probability of localising malfunctions caused by electromagnetic disturbances.

Description: Unless physical damage is caused by EMI, there is usually no evidence that it has occurred, other than a transient malfunction of the system, which may not even be noticed at the time. Physical damage caused by EMI is also likely to be misdiagnosed unless EMI detection is used to correlate events.

An event data recorder (EDR) can be used to enable the establishment of evidence that a malfunction, which could have been caused by EMI, has occurred.

Whenever an anomaly is detected (such as an out-of-range data value, checksum failure, sequencing error, etc.) relevant data can be recorded. For example, electromagnetically diverse software may reveal implementation errors via the discrepancy of results during operation, so all such discrepancies shall be timestamped and logged in an EDR when one is required.

This data can then be analysed statistically in real time or at some later time to detect and diagnose trends due to sporadic failures and to propose remedial action.

Data captured by an EDR can only reflect the events and malfunctions it has been designed and programmed to detect and record. Consequently, to be practically useful, an EDR must store information for the sort of event types adequate for diagnosing the system behaviour retrospectively.

To aid fault attribution and diagnostics the fault detection should be time-correlated to independent EMI event detection (see Section 2.2.9).

Identification: A routine can be called each time a malfunction is detected and should usually record, at the very least, the data itself and a time stamp code.

It is necessary for the resolution of the data recorded and its sample rate to be adequate for meaningful subsequent analysis.

Depending on the type of event recorder used and its mode of operation, pre-event data settings may also be important.

Depending on the size of the system and the safety integrity level/ systematic capability, the EDR might be physically separate (and able to be 'arrested' by the relevant safety authorities) for example, for a train or plane.

Mitigation: Analysis and diagnosis of the data can be used to look for co-related events and trends.

Future designs, or modifications to the existing design, should take the resulting information into account to keep pace with the worsening of the electromagnetic environment, and also to improve the risk-reduction achieved when using electromagnetic resilience techniques and measures.

Note 1: Also see the anti-tampering techniques and measures in Section 2.2.10.

Note 2: Consideration should also be given to increasing the electromagnetic immunity and/or electromagnetic resilience of the EDR, for example, by using techniques detailed in this Code of Practice, to ensure that an electromagnetic event that affects the safety-related system does not also affect the data stored in the EDR.

2.2.6 Improving the electromagnetic resilience of communication links

Aim/ description:
The electromagnetic resilience of a safety-related system can be made more robust by improving the electromagnetic resilience of its communication links, such as networks (for example, CAN, Profibus, Ethernet, wireless links including wide/local area networks, etc.), backplanes (for example, VME), printed circuit boards (ground planes) and even on-chip interconnect, by applying hardware and software techniques and measures.

Identification/ mitigation:
Hardware and software techniques should be used, either individually or together, to improve the reliability of the links.

Suitable hardware techniques are described in this Code of Practice. Suitable software techniques include, but are not limited to, those set out in Sections 2.2.6.1 - 3.

Wireless links are especially susceptible to electromagnetic disturbances (see Section 2.3.25).

References:
[100]-[102].

2.2.6.1 Error detection on parallel or serial buses

Description:
Redundant data is appended to the actual data using error detection coding (EDC) and error correction coding (ECC) techniques (for examples, see Sections 2.3.11 - 2.3.13).

This enables the detection of data corruption using techniques such as parity or cyclic redundancy checking (CRC).

Once data corruption is detected, appropriate action can be taken to maintain the safety integrity level/systematic capability, as described in the safety documentation. For example, various retry schemes could be used to improve the reliability of the link (at the expense of the overall system performance).

Where the safety manual for a sub-system or element includes a DS, it shall provide sufficient detail on it to allow its correct use by a safety system's designer.

2.2.6.2 Error correction on serial or parallel buses

Description:
This is a variation of the previous technique, however, the code is such that a level of error correction is possible in order to both detect corruption and also correct for its effects.

Various error correcting code (ECC) schemes (see Sections 2.3.11 - 2.3.13) can be used to improve the reliability of the link at the expense of a reduced data rate.

Whenever error correction occurs, this should be logged to aid later diagnosis (see Section 2.2.5).

2.2.6.3 Protection of a sequence

Description: When there is a stream of data packets on a data bus or communications link the packets may be duplicated, corrupted, delayed or lost during transmission possibly due to EMI.

Extra sequence codes can be appended to each packet to enable detection of delayed, lost or duplicated packets.

Various techniques and measures in this Code of Practice can be used at the packet level, for example, even just a single bit can be alternated between packets to detect a single packet failure (omission or duplication) (for example, see [107]).

More elaborate techniques are needed to detect multiple packet failures or corruption.

Identification: Depends on the technique used for marking the sequence of the packets.

2.2.7 Synchronisation and resynchronisation techniques

Aim: To improve the availability of a synchronous function or system in the event of a detected EMI-induced corruption.

Description: The ability of a synchronous function or system to detect that it is running abnormally and then reset its own state, or the state of the system, whilst maintaining its safety integrity level or systematic capability.

For example, in some processor architectures EMI can cause a processing exception due to corrupt data or the incorrect execution of an instruction.

Identification: By any appropriate techniques and measures, such as those described in this Code of Practice.

Mitigation: A clear and understandable system design concept is needed for the credible and practical implementation of this technique.

Different techniques may be needed to resynchronise continuous and non-continuous synchronous systems.

The application must be able to safely tolerate the reset or resynchronisation.

The use of low-level programming features may be necessary to implement state resynchronisation, or to return the system to a safe state.

The use of built-in exception handling (for example, https://en.wikipedia.org/wiki/Exception_handling) within the language runtime package or operating system should only be relied upon if the resulting response is deterministic and accommodated as part of the overall design.

The use of an electromagnetically diverse monitor should be considered (see Section 2.2.3).

Note: The Importance of this technique depends on whether the safety function is intended for: continuous operation; to operate 'on demand'; or where any kind of system has no safe state.

2.2.8 Protection from persistent interference by monitoring retry counts

Aim:

To improve system resilience during persistent failures including those caused by EMI.

If a system is exposed to persistent electromagnetic disturbances to which it is susceptible, causing it to suffer EMI, then the operation of the system may be severely affected or even halted.

For example, a communication link, even with a retry facility, may be so affected that no message traffic can successfully be communicated.

Any defence mechanism relying on reactivation of a function or retransmission of a message might be so affected that there is effectively a 'denial of service' (DoS), which may or may not be deliberate.

Description:

A task that continually monitors the retry counter values and timestamps of functions, memory checkers, communication protocols, and any other function that uses a retry or state recovery approach, to improve its perceived short term reliability.

This task itself would require some check for 'Liveness' [128], for example, a timeout in order to be effective, preferably based on an electromagnetically diverse independent hardware watchdog timer (see Sections 2.2.3 and 2.3.19).

Mitigation:

Possible approaches might be to switch to a backup system; to switch to manual operation; or to provide information to operators or maintainers. Many other possibilities for the end-use application should be considered at the system design stage.

Reference [143] may also be of use in the case of near-continuous electromagnetic interference.

Note:

The importance of this technique depends on whether the safety function is intended for continuous or on-demand operation.

2.2.9 Independent detection of electromagnetic disturbances and/or EMI

Aim: Detect electromagnetic disturbances in the environment and/or EMI in the equipment.

Description: Independent detectors are used to sense the occurrence of certain types (ideally, all types) of electromagnetic disturbances, although perhaps only when they exceed certain levels. An example is described in [664], and [665] describes current experience of a deployed IEMI detector. Several other types of detector have been developed, usually by military/security organisations.

Where certain electromagnetic signals are required for safe operation, such as GPS signals, some means to detect their absence or 'jamming' may be necessary for maintenance of the safety integrity level/systematic capability. (Also see Section 2.2.10, where the communication link is electromagnetically based).

An approach that relies on the internal resources of commercial off-the-shelf (COTS) devices operating system logs and other internal data and signals to provide valuable information about whether EMI is being experienced, is introduced in references [666] and [667]. An effective set of sensors has been identified for computers and smartphones and it has been shown that these observables were responsive to electromagnetic disturbances.

This definition of observables is empirical as it involves only resources that are accessible to users with simple or administrative rights in the operating system and which were not designed by the COTS manufacturers to be used for EMC testing or functional safety. Consequently, this approach could be improved by having CPU manufacturers and operating system editors provide more interfaces to gather low-level information about the health status of the system. This approach has the benefits of allowing the design of a real-time remote monitoring system for electromagnetic disturbances that cause EMI.

Mitigation: This technique may be used in many different ways, for example:

(a) to help manage the external conducted and/or radiated electromagnetic environment over the lifecycle, for example, by displaying or sounding a warning — or initiating other actions according to the safety documentation — if the equipment starts to experience levels of electromagnetic disturbance in excess of the level of immunity the equipment was designed to withstand.

It could, for example, warn of the use of equipment using high RF power, such as a diathermic heater, in too-close proximity. This technique has been used in hospitals to help enforce their 'no cellphones' policies by sounding a warning, and could be helpful in enforcing the walkie-talkie example in Section 2.3.4.

(b) by detecting a failure of electrostatic control measures (such as humidity control, static floor re-treatments, etc.) that could expose equipment to higher levels of ESD than it was designed to be able to cope with.

(The usual maximum ESD test level in immunity standards is ±8 kV, but levels of ±25 kV or more have been seen during reduced atmospheric humidity, and the automobile industry has tested to such levels for decades for this reason.)

(c) by making sure that certain sensor or transducer readings were ignored, or certain circuits were reset, for the duration of an excessive disturbance.

This is a well-established technique for preventing intentional interference with machines that can pay out money, for example gambling machines, change machines, automatic teller machines (ATMs), etc. (A typical tool used for such IEMI is the cattle prod, which generates impulses of around 35 kV.)

It has also been used with some very sensitive medical diagnostic instruments to warn when their results should be ignored because the electromagnetic environment was noisier than they were designed to cope with (sometimes at quite low levels, such as > 1 V/m).

(d) by recording data on the occurrence of certain types and/or levels of electromagnetic disturbances in an EDR (see Section 2.2.5), ideally with time-of-event correlation to help attribute and diagnose the causes of failures, after the fact.

(e) by monitoring the internal electromagnetic environment of equipment that relies on external shielding, filtering and/or surge protection so that if any of them should degrade, and if that degradation permits higher-than-acceptable levels of electromagnetic disturbance to enter the equipment, then action in accordance with the safety documentation can be initiated.

This could be helpful in enforcing Section 2.6.2 so that, for example, if someone uses an incorrect type of shielded cable, or does not terminate it correctly, an alarm is sounded.

2.2.10   Protection of systems from tampering via communication links to external systems

Aim: To conserve the safety integrity/systematic capability of systems, sub-systems or elements that have external communication links, especially with the internet, at least as regards electromagnetic resilience.

Description: Many systems are connected to the internet or an intranet and as such are vulnerable to hacking attacks, virus infestation, Trojan attacks, spoofing (imitation of identity), and DoS attacks.

The offensive techniques can be used to access, change or delete electronic data recorder (EDR) records and to change programs to make them more vulnerable to EMI.

Identification: Typically a firewall is used to prevent attack and enable protection of the EMI log, and keeps a record of the attacks it has detected, together with any consequent actions.

For EDRs that are built in, the removal and replacement record can be consulted.

Remember that it is the system integrator who is responsible for protecting the system from this kind of threat (see Section 2.1.4).

Mitigation: At least provide some protection of the EMI log by using a firewall to help prevent attacks from succeeding. Actually detecting and subsequently attributing a malicious event is more likely to be effective in a broader context than just achieving electromagnetic resilience.

If the EDR log media is physically removable then the records of its removal and replacement should be stored in non-volatile memory, which is built permanently into the system.

In the event of the EMI log being tampered with this record can be consulted.

Some EDR logs are built into the system and accessed interactively via a port. In this case it is necessary to restrict access to 'read only' so that the EDR data cannot be altered or deleted, thus destroying possible evidence.

EDR data may be encrypted to make tampering harder and alteration easier to detect.

2.2.11 Robust, high-specification electromagnetic mitigation

Aim: To provide a benign 'internal electromagnetic environment' by reliably attenuating the external electromagnetic environment to a very high degree, over the anticipated lifecycle.

Description: A combination of high-specification electromagnetic mitigation including shielding, filtering, transient suppression, galvanic isolation, etc., traditionally taking the form of a mechanically rugged metal enclosure fitted with bulkhead-mounted cable connectors incorporating robust filtering, transient suppression and/or galvanic isolation.

This combination is designed so as to provide reliable attenuation of all electromagnetic disturbances, possibly even including direct lightning strike and electromagnetic pulse (EMP - see Section 2.1.4), over the entire lifecycle by a suitable combination of initial design plus regular maintenance, repair and refurbishment, which includes reverification of mitigation performance.

EM detection techniques (see Section 2.2.9) might be able to be used within an overall enclosure used for this purpose, in order to provide prior indication of certain failures or degradations in mitigation, perhaps enabling repair and refurbishment to take place when needed outside of the regular maintenance schedule. This approach can also be useful to help identify foreseeable misuse, such as doors or panels left open or not fitted properly, the use of incorrect types of cables/connectors, etc., or to identify electromagnetic disturbances that exceed those covered by the original design.

Robust, high-specification electromagnetic mitigation, when implemented correctly, can allow an electronic system to operate continuously throughout any/all external electromagnetic disturbances, so can be very useful when degradation or interruption of functionality is not desired.

Identification/ mitigation: With appropriate design, this technique can be used to address any external (i.e. inter-system) electromagnetic disturbances over the anticipated lifecycle.

However, it cannot deal with intra-system (internal) electromagnetic disturbances.

Note: The extent to which robust conventional EMC mitigation techniques (for example, high-specification electromagnetic mitigation including shielding, filtering, transient suppression, galvanic isolation, etc.) can prevent electromagnetic disturbances from affecting the correct operation of a safety-related system during its anticipated lifecycle may be taken into account during the selection and application of the techniques and measures, where this is justified.

2.3 Techniques and measures for use in operational design

When the design is implemented the functionality may be realized in hardware and/or software. In the subsections below, techniques and measures are classified as either hardware or software based, but some techniques and measures may have equivalent representations in either hardware or software, which might be more effective.

Note: Where a technique or measure in this section applies to a technology that is not relevant to the equipment or system concerned, and the importance as shown in Annex A as being M or HR, a justification for why that technique or measure was not applied should be included in the safety documentation (see Section 1.8).

2.3.1 Developing appropriate operation and maintenance instructions

Aim: To develop instructions for procedures that help to avoid EMI-induced failures during the operation and maintenance of a safety-related system or a sub-system or element used within a safety-related system.

Description: This is where the operation and maintenance requirements – and their justifications – for the electromagnetic resilience techniques and methods used to comply with the design specification are documented (also see Clause 7.6 of [119]).

The operation instructions may include, for example:

(a) restrictions on the use of potentially interfering equipment in the vicinity of the safety system (such as mobile phones, cellphones, welding equipment etc.).

(b) restrictions on the removal of access panels where these contribute to protection from electromagnetic disturbances.

(c) for portable safety-related equipment, restrictions on the type of electromagnetic environment in which the equipment is intended to be used.

(d) restrictions in the use of the safety-related equipment, for example, where it is user-configurable, where this may affect protection from electromagnetic disturbances.

(e) requirements for recording and reporting system upsets, system restarts, safe failures, trips to safe state etc., especially where the cause is not obvious and may be due to an EMI event. (Recording and assessing system trips is an important contributor to reliability growth in general, and could be the only indication that the electromagnetic protection is not operating as intended.)

(f) requirements for monitoring the electromagnetic environment and detecting/recording EMI events to enable correlation with faults.

The maintenance instructions may include, for example:

(g) monitoring/inspection of physical protection measures against electromagnetic disturbances, such as access panel/door gaskets for deterioration or corrosion of mating surfaces, shielding effectiveness, etc.

(h) recommendations on the inspection and maintenance intervals necessary to maintain physical defences against electromagnetic disturbances.

(i) any lifetime restrictions due to the anticipated degradation of physical protection measures against electromagnetic disturbances, such as those due to corrosion.

(j) procedures to be followed to verify the continued effectiveness of physical protection measures after an unusual electromagnetic disturbance event, such as a major power surge, nearby lightning strike, etc.

Identification: By independent assessment of the relevant documents against the guidance in this Code of Practice; see Clause 8 of [8] for guidance on the appropriate level of independence.

Mitigation: By correction of the relevant documents.

Note: Experience indicates that operation and maintenance instructions should only be expected to achieve a risk reduction factor of no more than two.

2.3.2 Designing appropriate maintenance techniques

Aim: The design should make it practical to monitor the condition/performance of, and replace if necessary, electromagnetic mitigation items such as filters, surge suppressors, conductive gaskets, etc., which may have a limited operational life.

Identification: By independent assessment of how easy it is for the relevant people to monitor and replace electromagnetic mitigation items that may have a limited life.

Mitigation: By correction of the relevant documents.

Note: Experience indicates that maintenance instructions should only be expected to achieve a risk reduction factor of no more than two.

2.3.3 Limiting the possibilities for operation and hence mis-operation

Aim: EMI can affect operator controls, creating the same effect as an unskilled or even malicious operator. This technique helps to avoid operation in unwanted or unnecessary modes.

Description: This approach reduces the operation possibilities, and therefore the possibilities for EMI to cause failures, by limiting, for example:

(a) the number of generally possible operating modes;

(b) physically protected operation of special operating modes, for example, by using key switches that are lockable or have protected access;

(c) the number of operating elements; and

(d) consistency checks specifically aimed at detecting operationally inconsistent or non-plausible operating modes.

The hardware and/or software design techniques and measures used for limiting the possibilities for operation should comply with the requirements of this Code of Practice.

Identification: Competent independent assessment of the hardware and/or software design techniques used for limiting the possibilities for operation.

Mitigation: By modification of the design using appropriate techniques, for example, those techniques described in this Code of Practice.

2.3.4 Protecting against operation errors

Aim: To protect the system against operator errors, mistakes and other foreseeable misuse.

Description: Incorrect operator inputs (value, time, etc.) are detected via plausibility checks, monitoring of the EUC or other methods.

To integrate these facilities into the design, it is necessary to state at a very early stage which inputs are possible and which are permissible.

A mistake in operation should not result in dangerous failure. Such foreseeable use/misuse should never be permitted to compromise functional safety.

Identification: Competent independent assessment of the hardware and/or software design techniques and measures used for the protection against operator mistakes.

Mitigation: By hazard analysis, modification of the design and logging of mal-operations, using appropriate techniques and measures, for example, those described in this Code of Practice.

> For example, using a walkie-talkie or cellphone closer than is permitted, or the failure to correctly close a shielding door, or to refit a shielding inspection panel, could reduce availability and/or prevent the attainment of a safe state (see Section 2.2.9).

2.3.5 Protecting against hardware or software modifications or manipulations

Aim: To protect the safety-related system against hardware or software modifications or manipulations by any technical means.

Description: Modifications or technical manipulations are detected automatically, for example, by plausibility checks for the sensor signals, detection by the technical process, automatic start-up tests, etc.

If an unapproved modification or technical manipulation is detected, appropriate action is taken in accordance with the safety documentation.

(Section 2.2.9 describes one way of detecting modifications that could degrade electromagnetic mitigation.)

Identification: Competent independent assessment of the hardware and/or software design techniques used for detecting modifications or manipulations.

Mitigation: By modification of the design, using techniques and measures that comply with the requirements of this Code of Practice.

Note: Modifications should be subject to a documented change control procedure and should not compromise the safety documentation or functional safety.

2.3.6 Defensive programming techniques

Aim: To design software programs in such a way that they will detect anomalous control flow, data flow or data values that may have been caused by EMI during their execution and to react in a predetermined and acceptable manner.

Description: Many techniques can be used during programming to detect and control the anomalies induced by EMI-induced corruption; see the references.

A range of error detection and/or correction techniques and measures, such as those described in this Code of Practice, can be used to implement an acceptable hardware/software solution.

To aid fault attribution and diagnostics the fault detection should be correlated to independent EMI event detection (see Section 2.2.9).

Identification/ mitigation: The principal defensive mechanisms are listed below, in Sections 2.3.6.1 - 2.3.6.3.

Where the safety manual for a sub-system or element includes a DS, it shall provide sufficient detail on it to allow its correct use by a safety system's designer.

References: [100]-[102] and [104] and by prevention: [113]-[116].

2.3.6.1 Range checking in hardware and in software

Identification: Range checking of the values of all variables, for credibility.

This is achieved by defining a number of bands for the value of each variable, the meaning of the bands being specific to the application.

A typical example of three bands is: normal operational values; warning zone values; and out of range values.

This applies to values anywhere in the processing chain, not just I/O 'signals', whether they are analogue or digital. It also applies to algorithms implanted in hardware such as FPGAs.

This is valuable for EMI detection as the value of the original variable may have been corrupted by an EMI event.

A program might well be correct but the result of an assignment might be 'out of range' and cause the program to malfunction.

The programming language provides a means of assigning a data type to a data variable to define the range (or set) of values that it is intended to contain.

Whenever values are assigned to the variable, either at compile time (constant values) or at runtime (constant or modified values) then a check is made that the new data value is within the range of values specified by the type of the variable.

IEC 61508 calls this 'strong data typing'.

In any case all variables should be initialised explicitly to an acceptable value, before being used, so that out-of-range errors are not caused by the arbitrary value in memory when power is first applied.

Mitigation: If the language's run-time package supports range checking, then that can be used (bearing in mind the loss of performance and increased size of program). If there is no automatic run-time range checking, then explicit tests should be designed into the program. This also applies to hardware, for example hardware specified algorithmically in languages such as Verilog, including in FPGAs.

Range checks can be implemented both at:

(a) 'Compile time', using assertions about the value ranges that the software/logic is specified to handle. This is often referred to as 'static analysis'; it has no runtime overhead.
(b) 'Operation time' or 'run time', using program checks during system operation to verify values before they are relied upon for decision-making.

This is often referred to as 'dynamic analysis'. The run-time load must be allowed for in the system design.

References: [100], [102] and [104].

2.3.6.2 Sequence checking

Identification: Sequence checking is a powerful technique for ensuring that a sequence of values or stream of data packets is in the correct order and that there is no duplication or omissions.

Sequence checking can be used for data and also for program state, for example, using finite state machines or Petri nets.

The program contains intermediate points where the expected state of the program, i.e. the values of data or status variables, can be checked for credibility.

Mitigation:	Various techniques and measures, such as those described in this Code of Practice, can be used at the hardware level to implement an acceptable solution.
	Communication protocols using sequencing can be used to improve the effective quality and reliability of the link, for example, packet sequencing.
	If the program is detected as being out of sequence then this fact can be logged and then, if appropriate, a recovery attempted so that processing can continue from a known valid state.
References:	[102], [104] and [107].
Note:	The importance of this technique depends on whether the safety function is intended for continuous operation or on demand.

2.3.6.3   Correct rounding and resolution in all calculations

Description:	The incorrect handling of rounding errors and resolution (fixed or floating point) has been the cause of many high profile project failures, such as Ariane 5, see [129], and the Mars probe, see [130]. Where different parts of systems use different units of measurement then conversions between data used need to be carefully checked in all contexts. The corruption of data by EMI may cause invalid values of data to occur and software exception handling techniques, such as range checking, can be used to verify the plausibility of data before it is relied upon by a safety critical function.
References:	[120], [121], [129] (an example of poor exception handling), [130] (an example of incompatible units), [141].

2.3.7   Limited use of interrupts

Aim:	To reduce the likelihood that EM disturbances will affect the execution of the software.
	EMI can increase the likelihood that spurious interrupts are generated, possibly at such high rates that the timing of the software execution can be affected.
	Interrupts can arrive asynchronously and, of course, interrupt the flow of the main program and possibly other interrupt routines that may be running at the time.
	Interrupts are therefore prone to causing errors, and the determinism of the program's behaviour becomes very difficult to predict. For example, can it be guaranteed that an interrupt routine will never cause a loop that freezes the whole system?
Description:	The use of interrupts should be restricted, but they may be used if they simplify the safety-related system to give an overall advantage for functional safety.
	It is understood that some very critical nuclear and military software is designed without any interrupts at all, in order to improve the determinism of the program's behaviour.
Identification:	At compile time a static analysis program may be used to flag up any use of interrupts.

Mitigation:	The use of interrupt routines should be limited and, when used, their effect on system timing and the sharing of computing resources should be documented.
	Software handling of interrupts should be inhibited during critical parts (for example, time critical, critical-to-data changes) of the executed functions.
	If interrupts are used, then parts not interruptible should have a specified maximum computation time, so that the maximum time for which an interrupt is inhibited can be calculated.
	Also see Section 2.2.1.
References:	[123], [125], [144] and [145].

2.3.8 Limited use of memory address pointer variables

Aim:	To reduce the impact of memory corruption due to EMI.
Description:	A pointer is a variable with a value that is an address of data in memory.
	If the pointer variable is corrupted by EMI then the impact on the behaviour of the program is likely to be unpredictable. For example, the corrupted pointer may either be pointing at some data, the program subroutine stack, the heap, or even the program itself, and consequently any write operation will corrupt the system.
Identification:	At compile time a static analysis program may be used to flag up any use of pointers.
Mitigation:	A set of programming guidelines would normally prohibit the explicit use of pointers, unless this is essential from an algorithmic viewpoint and its use can be clearly justified in the safety documentation.
	If the hardware or run-time system architecture allows memory address ranges to have protected access then this feature can be used to ensure that only the intended memory partitions are accessible in each context. This would also make available the means for detecting an access violation. However, it would not detect data content corruption within accessible address ranges.
	Partitioned ranges of memory and/or a memory management unit can be used to detect violations and provide some measure of protection (see Section 2.3.10.3).
References:	[113]-[116].
Note:	The importance of this technique depends on whether the safety function is intended for: continuous operation; to operate 'on demand'; or where any kind of system has no safe state.

2.3.9 Avoiding recursion

Aim: To help reduce the impact of corruption due to EMI on program execution.

Description: Recursion is the act of a program calling or referencing a part of itself, either directly or indirectly.

It is more susceptible to the effects of EMI-induced corruption as the nested chain of calls is held as a linked list of pointers on the stack, in effect potentially a very large list of pointers that increases susceptibility to electromagnetic disturbances. The deeper the level of recursion used the more susceptible the implementation. In general the use of recursion can be replaced by an equivalent loop structure; this avoids extensive use of pointers and the possibility of running out of memory used to accommodate the pointer linkages for implementing recursion.

Recursion should only be used with the greatest caution – and comprehensive justification in the safety documentation – in safety-related software.

Identification: At compile time a static analysis program may be used to find instances of recursion in the program source text.

Mitigation: Programming guidelines would normally prohibit the use of recursion unless its use is fully analysed for resource usage and is clearly justified in the safety documentation. This would require a rigorous argument for, or proof of, the maximum depth of recursion that would be experienced during operation, and the amount of memory that would be required to support this at runtime.

Every algorithm that can be expressed using recursion also has an equivalent using an iterative looping construct. In general the latter should be the preferred solution for safety-related systems or equipment intended to be used in them.

2.3.10 Error detection and correction for invariable memory

Overall aim: The detection of information modifications in the invariable memory (i.e. ROM, or program memory).

Overall mitigation: Techniques and measures should be applied accordingly, bearing in mind all the possible susceptibilities of the system to the variety of electromagnetic disturbances described in Section 2.1.

Some other suitable techniques and measures are described in Section 2.3, and alternative or additional techniques may be used if technical justifications for them are provided in the safety documentation.

2.3.10.1 Signature of a word or block of data

Aim: To detect single and multi-bit corruption within a block of data. Various checking techniques are available, such as cyclic redundancy checks (CRC), secure hash algorithm (SHA), and Hamming codes (for correction as well as detection).

Description: This procedure calculates a signature using an error-checking technique. The extended signature is stored, recalculated and compared as in the single-word case. A failure is indicated if there is a difference between the stored and recalculated signatures.

Identification/ Mitigation:	When an error is detected, apply a response defined by the safety documentation. The error detection and/or correction method used should be commensurate with the requirements of the safety integrity level/systematic capability.
	Where a safety manual for a sub-system or element includes a DS, it shall provide sufficient detail on it to allow its correct use by a safety system's designer.
References:	[108]-[112] inclusive, [131] for Hamming codes and CRC, [132] for SHA.

2.3.10.2 Block replication with inversion (for example, dual redundant ROM with comparison)

Aim:	To detect all bit failures.
	This is a powerful technique that should be used wherever practicable.
Description:	The address space is duplicated in two memories, which ideally should be physically separate. The data is stored inversely in one of the two memories and inverted again to be compared with the other copy.
	The inversion of the data in one memory makes this technique much more effective against the common-cause errors, malfunctions or failures including the typical effects of EMI.
Identification:	The outputs are compared and a failure indication is produced if a difference is detected.
Mitigation:	Repeat the memory read as many times as necessary without unacceptably degrading the safety integrity. If the failure clears, continue operation as usual. In any case, if a log is available, the fault should be recorded (see Section 2.2.5). If during the time available the failure does not disappear, apply an appropriate response defined by the safety documentation.
	Where a safety manual for a sub-system or element includes a DS, it shall provide sufficient detail on it to allow its correct use by a safety system's designer.
Note:	The use of electromagnetically diverse memories improves the effectiveness of this technique for electromagnetic resilience (see Section 2.2.3).

2.3.10.3 Memory boundary protection

Aim:	Memory boundary protection, by providing access-protection boundaries between separate memory areas within a program, and also between separate programs running on the same processor(s).
Description:	Runtime plausibility checking of use of a memory segmented into partitions. This is important as EMI-induced corruption of the program counter, stack pointer, heap pointer or any pointer in a program could cause data to be written to a wrong memory address, resulting in corruption of data or execution of the storing program instructions.
	Statically defined and protected address ranges are used for the following:

    (a) program;
    (b) stack;
    (c) statically-allocated variables;
    (d) heap (dynamically allocated variables);
    (e) inputs; and
    (f) outputs.

Identification: This technique simply prevents incorrect memory areas from being used, for example, by the effects of EMI on the address bus.

If the mechanism used to manage memory accesses can detect out-of-range addressing violations, they could be logged to support testing and diagnosis of system malfunction.

Mitigation: Upon detection, apply an appropriate response defined in the safety documentation.

Where a safety manual for a sub-system or element includes a DS, it shall provide sufficient detail on it to allow its correct use by a safety system's designer.

References: [113]-[116].

2.3.11 Error detection and correction techniques in redundant designs

Aim: To enhance electromagnetic resilience by comparing the results of multiple redundant channels in hardware or software.

Description: The system may be replicated using one or more processors and/or buses. Each system independently determines the next action to be taken and their results are compared before the action is sanctioned. Various schemes can be used, for example, two channels, three channels, one channel per processor or multiple channels per processor.

Where duplicate or triplicate channels are used without hardware diversity, with or without software diversity, the effectiveness of this technique against common-cause errors can be increased by ensuring that the channels are desynchronised, or if synchronous are kept out of step with one another. This makes it less likely that EMI will affect all the channels in the same way.

Similarly, to increase the effectiveness of this technique against the common-cause errors, malfunctions or failures typical of EMI, electromagnetically diverse encoding of data and or programs can be used (see Section 2.2.3).

When multiple channels are implemented on physically separate processors the resilience will be enhanced if the power supplies are isolated and the interconnections are properly protected against electromagnetic disturbances.

Identification: The result of comparing the sets of signals must be acceptable for safety in the current context.

The comparator or voter (the circuit used to compare channels and detect errors) is a potential single point of failure and so must be designed to have considerably greater resilience to electromagnetic disturbances for this technique to be effective. This may be achieved by, for example, the strong use of self-testing to verify correct operation, switching to a redundant comparator or voter (ideally one using diverse technology; see Section 2.2.9) if necessary. An alternative possibility is the use of rugged, lifetime-reliable, high-specification electromagnetic mitigation, which may be achievable with small size and low weight on a printed circuit board (see [69]).

Mitigation: Upon detection of an anomaly, apply an appropriate response as defined in the safety documentation.

Where a safety manual for a sub-system or element includes a DS, it shall provide sufficient detail on it to allow its correct use by a safety system's designer.

Note: The importance of this technique depends on whether the safety function is intended for continuous operation or on demand.

2.3.12 Time-based error detection/correction in buses and interfaces

Aim: To detect transient failures in bus and/or interface communication.

Description: The information is transferred several times in sequence.

                        The repetition is effective only against transient failures.

Identification: Each instance of the information is stored as it is received and then the instances are compared to see if they are consistent.

                        Often sequence numbers or time stamps are incorporated into the data so that it becomes possible to check that data has arrived in the correct order and that none have been lost in transit.

                        To improve the effectiveness of this technique it is often combined with the use of error-checking codes to protect the sequence numbers or time codes (see Sections 2.3.11 and 2.3.13).

Mitigation: Apply an appropriate response as defined in the safety documentation.

                        Where a safety manual for a sub-system or element includes a DS, it shall provide sufficient detail on it to allow its correct use by a safety system's designer.

Note: Requires at least one complete repetition in one cycle time of the process.

References: [100], [102], [104], [107] and [109].

2.3.13 Error detection and correction for variable memory

Global aim: Detecting failures during addressing, writing, storing and reading data in memory.

2.3.13.1 Memory testing

Aim: To provide memory testing before operation and/or during operation to detect errors specific to memory systems.

Description:

It is crucial that read/write memory devices function correctly in order for any computer-based system to work reliably. The content of memory devices can be corrupted by EMI and the devices themselves might even be physically damaged by severe EMI.

It is therefore necessary to efficiently test the memory before operation and during operation to ensure that it is functioning normally. It is also necessary to design the testing in such a way that the known causes of error are tested as separately as possible.

This becomes more and more important as products and systems increase their use and size of memory. It is important that memory tests do not corrupt the running of the system itself, for example, the content of memory stacks, heaps and configuration data.

However, even well-designed memory systems are naturally susceptible to EMI, in particular when they rely on the storage of electrical charge to represent digital values as charge can be altered by EMI (and by ionising radiation).

The purpose of the memory test is to ensure that the memory device is fully functional, so the test needs to write a set of data to each individual address in the device and verify its correct value by reading the data back. In cases where such a test would destroy the data in the memory precautions need to be taken, such as copying it, or specific techniques are used to conserve the memory content in situ.

The tests used need to be designed to efficiently identify the likely kinds of faults that the memory device being tested is likely to suffer from, or have induced by, electromagnetic disturbances and/or ionising radiation.

Catastrophic internal failure of devices sometimes occurs and must be detectable, but most memory failures are caused by wiring problems, including crosstalk on the data, address and control line busses. Often these problems are difficult to detect and isolate as the memory affected may not actually be used for extended periods of time. When it is used and causes mal-operation of the system it can be very difficult to diagnose and isolate the originating cause.

A memory test strategy that takes account of at least the following points is required:

(a) detection of missing memory chips;

(b) detection of incompletely inserted or connected memory devices;

(c) testing the memory data bus, preferably one bit at a time to aid fault isolation;

(d) testing the memory address bus, primarily to detect that address bus faults are not causing overlapping memory locations to be addressed;

(e) testing the function of the memory device itself;

(f) if dynamic memories are being tested then signal integrity tests for DDR2/DDR3 memory signal lines would also need to be tested;

(g) testing of systems using 'memory caching' techniques requires special consideration, such as provision for the safe flushing of cache memory, which might otherwise temporarily mask a fault that has developed after the saved data has been cached; and

(h) testing of memory currently in use by the system (its stacks, heaps and state variables) requires great care in order for the Testing not to corrupt the system itself. This is of particular concern for operational systems when memory tests are periodically scheduled as background activities to check on-going system integrity.

Detection:

The memory test can be run:

(a) during the initialisation of the system and after any reset of the system; i.e. before operational use of the system starts; and/or

(b) during the real-time operation of the system.

Mitigation: Analysis and diagnosis of the memory test result data can be used to identify specific kinds of common faults in the memory system efficiently.

Future designs, or modifications to the existing design, should take the resulting information into account to keep pace with the changes in memory technology, in particular its susceptibility to EMI as memory cell sizes continue to be reduced, compounded by the worsening of the electromagnetic environment.

Note: Also apply other electromagnetic resilience techniques and measures as appropriate for testing memories and achieving the aims of this section.

References: [133]-[136].

2.3.13.2 One-bit redundancy

Aim: To detect some changes in the content of a memory location, bus or I/O register.

Description: Every data word is extended by a single bit, often called the parity bit, based on the binary value of the data.

Identification: The parity bit of the data word is set when it is stored, and then checked each time it is read.

If an invalid parity value is detected it indicates that the content has been corrupted. A failure action can then be activated.

If the parity value is correct then there may have been no error, however, there may have been multiple bit changes to the content resulting in the same parity value.

Mitigation: Upon detection, apply an appropriate response as defined in the safety documentation.

Where a safety manual for a sub-system or element includes a DS, it shall provide sufficient detail on it to allow its correct use by a safety system's designer.

References: [107]

2.3.13.3 Block replication with inversion to detect all bit failures

The techniques and measures in Section 2.3.10.2 for invariable memory also apply here.

2.3.13.4 Memory boundary protection

The techniques and measures in Section 2.3.10.3 for invariable memory also apply here.

2.3.14 Error detecting/correcting coding for ROM, RAM, buses and interfaces

Aim: To detect and/or correct one or more bit failures in a word.

Description: The memory, or the content of a data stream, is extended by one or more bits. Data code protection provides for dataflow-dependent failure detection, based on information redundancy (for example, CRC or Hamming codes) and/or time redundancy.

Identification: Every time data is handled, either hardware or software can determine whether a corruption has taken place by checking the additional bits.

The number of additional bits establishes the number of bit errors in the data word that can be detected and/or corrected.

Mitigation: If a difference is found, corrective action can be taken (or a failure indication produced) as defined in the safety documentation.

Where a safety manual for a sub-system or element includes a DS, it shall provide sufficient detail on it to allow its correct use by a safety system's designer.

Correction of the data can be used to maintain the correct operation of the safety function. The strength of the technique used shall be justified in the safety documentation.

References: [110]-[112]

2.3.15 Error detection and correction for logic and data processing

Global aim: To recognise any failures that could lead to incorrect results in processing units.

All the techniques and measures listed in this section are concerned with detecting failures in the processing units and soft failures (bit flips) in memories and registers, and are therefore useful for detecting damage caused by lightning (or other) surges and electrostatic discharges, as well as soft failures such as those caused by ionising radiation etc.

2.3.15.1 Self-test supported by hardware (one-channel)

Description: Additional special hardware supports self-test functions, for example, it monitors the output of a certain bit pattern, often referred to as a 'signature'. It is a form of watchdog that relies on data content rather than time.

Identification: Used for detecting disruption of program execution.

Coverage depends on the extent of the software functions generating the bit pattern signature.

Mitigation: Corrective action can be taken, or a failure indication produced, as defined in the safety documentation. Where a safety manual for a sub-system or element includes a DS, it shall provide sufficient detail on it to allow its correct use by a safety system's designer.

The additional hardware could, for example, drive a safety-related system to a safe state and/or restart it (if it is safe to do so).

It is usual for the additional hardware to be low-technology (i.e. not programmable or electromechanical), or even mechanical, pneumatic or hydraulic as they are not affected by electromagnetic disturbances at all.

2.3.15.2 Coded processing (one-channel)

Description: Processing unit designed with special failure-recognition or failure-correction circuit techniques.

Typically a detection mechanism, such as a watchdog timer, can be used to detect a malfunction affecting the safety of the system. In response the system may be reset to a known state, often referred to as a 'restore point', and the continuation of operation of the system attempted.

Identification: The detection of safety or operational malfunction shall be independent of the main processing system.

For example, a watchdog timer that is implemented by a separate piece of hardware in such a way as to be itself electromagnetically resilient, otherwise false system resets and restores would be triggered.

Mitigation: The system-level implications of resetting to each reachable restore point before continuation, at any time during system operation, need to be considered.

When used, the benefits to electromagnetic resilience should be assessed for the particular implementation, and the analysis recorded in the safety documentation.

Where a safety manual for a sub-system or element includes a DS, it shall provide sufficient detail on it to allow its correct use by a safety system's designer.

References: [100]-[102], [122] and [124].

2.3.15.3   Reciprocal comparison by software

Description: Two or more electromagnetically diverse processing units exchange data (results, intermediate results and test data) and cross-check at defined 'restore points' from which system operation could be continued in the event of a discrepancy. Detected differences indicate a failure.

Identification: Coverage of data discrepancies is high and detection can be fast.

Excellent against hard failures and can be good against soft and transient failures too.

Mitigation: If the diagnostic test interval is short compared to the process safety time, a restart may be possible while keeping the process running.

If the failed unit can be identified, continued operation with the healthy unit may be possible. Otherwise, the safety function must achieve a safe state.

Where a safety manual for a sub-system or element includes a DS, it shall provide sufficient detail on it to allow its correct use by a safety system's designer.

Note: Hardware and/or software diversity (see Section 2.2.3) shall be used to greatly improve coverage of the common-cause errors, malfunctions and failures typical of EMI.

2.3.15.4   Self-test by software during operation

Description: Standard processing unit hardware with additional software functions that run self-tests.

Identification: Can detect some failures but coverage is low. The self-test may also be affected by the failure.

Mitigation: May require additional monitoring circuitry to achieve a safe state on failure.

Where a safety manual for a sub-system or element includes a DS, it shall provide sufficient detail on it to allow its correct use by a safety system's designer.

References: [124] and Section 2.3.13.1.

2.3.16 Error detection and correction for electrical and electromechanical components

Aim: To control failures in electromechanical components, such as relays, actuators, magnetic logic devices etc.

Description: Electrical and electromechanical components are generally less susceptible to EMI-induced failure than electronic components as their operating signal levels are usually much higher, but they are never totally immune.

Direct failures due to gross overload causing contact welding or coil burnout are possible in some applications.

EMI to circuits that control electromechanical devices may cause failures due to:

(a) chatter (unintended repeated operation causing early wear-out);

(b) generation of additional electromagnetic disturbances via arcing or sparking at electrical contacts; or

(c) paralysis (device physically stuck).

Identification: Electromechanical components may be monitored as part of loop, for example, by relay contact monitoring, by actuator position monitoring, or by the effects on the EUC (on-line monitoring). Care should be taken that such monitoring will detect chatter (especially in relays) or partial operation in actuators.

The use of electromagnetically diverse technologies (see Section 2.2.3) is recommended when performing parallel functions to help deal with the common-cause effects of electromagnetic disturbances.

Mitigation: Burn-out or paralysis failures should be designed to achieve a safe state.

Where a safety manual for a sub-system or element includes a DS, it shall provide sufficient detail on it to allow its correct use by a safety system's designer.

Multi-channel systems may be able to tolerate a single-channel failure but the likelihood of common mode failures must be considered.

Examples are suppression of arcing and proper termination of inductive loads to avoid induced spikes.

2.3.17 Caution when using hardware or software libraries

Aim: To ensure that the electromagnetic resilience measures are applied and verified throughout the whole software design and implementation of a safety related system.

Description: Software development relies more and more on 'standard' components encapsulated in libraries, for example, a TCP-IP stack, a matrix multiplication package etc. They are the software equivalent of COTS hardware. The overall software is as strong as its weakest part and so it is essential for safety related systems that any library software used should be designed to the appropriate guidelines, such as those described in this Code of Practice.

Identification: The source code of all library code must be audited to ensure that electromagnetic resilience techniques have been competently and adequately applied.

Mitigation: Any hardware or software components or modules copied from any library must comply with this Code of Practice before being incorporated into an operational system.

Note: Hardware specification is now often implemented algorithmically (for example, Verilog) for FPGAs.

2.3.18 Error detection and correction for electronic components

Global aim: To control failures in solid-state active and passive components.

2.3.18.1 Tests by redundant hardware

Aim: To use additional hardware to monitor the operation of the relevant function (for example, a safety function, in a safety-related system).

Description: Redundant hardware can be used to provide diagnostic testing for safety functions.

Identification: Good for detecting failed states, but may be poor at detecting transient failures.

Coverage depends on the rate of test compared to the process periodicity.

Mitigation: Effectiveness depends on diagnostic coverage and diagnostic test interval compared to the process periodicity. If/when used, the benefits to electromagnetic resilience should be assessed for the particular implementation and the analysis recorded in the safety documentation.

Where a safety manual for a sub-system or element includes a DS, it shall provide sufficient detail on it to allow its correct use by a safety system's designer.

2.3.18.2 Using dynamic signalling techniques

Aim: To detect static failures by dynamic signal communications and processing.

Description: A forced change of otherwise static signals helps to detect static failures.

For example, alternating voltage signals are less vulnerable to stuck-at faults than static (direct voltage) signals.

Identification/ mitigation: Good at detecting failed states, but poor at detecting transient failures. If/when used, the benefits to electromagnetic resilience should be assessed for the particular implementation, and the analysis recorded in the safety documentation.

Where a safety manual for a sub-system or element includes a DS, it shall provide sufficient detail on it to allow its correct use by a safety system's designer.

2.3.18.3 Caution with use of test access ports and boundary-scan

Aim: To prevent any added test/diagnostics, especially boundary-scan (such as JTAG) from making the system more susceptible to electromagnetic disturbances.

Description: The added interconnections can make susceptibility worse, especially boundary scan (which adds logic between the I/O buffers and the IC's core logic to allow testing the core logic, creating a possible path for electromagnetic disturbances right into the 'heart' of any electronics).

All test access ports and their connections must be electromagnetically resilient to avoid the system being made more susceptible instead of less. Using a low-profile PCB-surface-mounted connector for the JTAG access port, and ensuring that no cable is ever left attached to it, is very helpful in achieving electromagnetic resilience.

Identification/ Mitigation: Where used, their effectiveness against electromagnetic disturbances should be determined taking into account the particular design features, and this analysis recorded in the safety documentation.

References: [137]-[138]

2.3.18.4 Monitored redundancy

Aim: To compare the behaviour of two or more channels in a multi-channel system, to detect errors and/or to correct them.

Description: The safety function is executed by at least two electromagnetically-diverse hardware channels (see Section 2.2.3). The outputs of these channels are monitored and if the output states differ a suitable action is initiated to maintain the safety integrity level/systematic capability.

Identification: Effective against static and transient failures, provided the monitoring system is not itself prone to EMI.

Mitigation: For a safety-related system: transition to a safe state as recorded in the safety documentation.

For a sub-system or element: transition to a DS as recorded in the safety manual, which shall provide sufficient detail on it to allow its correct use by a safety system's designer.

However, with three or more channels and a voting function, error correction (isolation of the faulty channel and continued operation) may be practicable (see Section 2.2.3).

2.3.18.5 Hardware with automatic self-test

Aim: To detect faults by periodic checking of the safety functions using automatic self-tests.

Description: The hardware tests itself repeatedly at suitable intervals.

Identification: Will only detect failed states, not the transient failures that may have caused them.

Mitigation: For a safety-related system: transition to a safe state as recorded in the safety documentation.

For a sub-system or element: transition to a DS as recorded in the safety manual, which shall provide sufficient detail on it to allow its correct use by a safety system's designer.

However, by using redundant electromagnetically-diverse-technology channels (see Section 2.2.3) it may be practicable to continue safe operation by switching from a failed channel to one that is still operating correctly.

2.3.18.6 Analogue signal monitoring

Aim: To improve confidence in signals and controls.

Description: Analogue signals are used in preference to digital on/off states.

Trip or safe states are represented by analogue signal levels, which can be continuously monitored for credibility (for example, by using window comparators for amplitude ranges; high-pass, band-pass and low-pass filters for frequency ranges, etc.).

Identification: Can be effective against EMI, especially if unusual signals are detected, logged and investigated.

Mitigation: Upon detection of an anomaly, apply an appropriate response as defined in the safety documentation.

For a safety-related system, transition to a safe state as recorded in the safety documentation.

For a sub-system or element: transition to a DS as recorded in the safety manual, which shall provide sufficient detail on it to allow its correct use by a safety system's designer.

However, by using redundant electromagnetically-diverse-technology channels (see Section 2.2.3) it may be practicable to continue safe operation by switching from a failed channel to one that is still operating correctly.

Signals can also be 'smoothed' in hardware and/or software up to the maximum permitted for the accuracy and responsiveness required.

Information on signal anomalies from event logs may be able to be used to improve long term electromagnetic resilience and future designs.

2.3.18.7 'Data assurance' (content credibility checking)

Aim: To use known relationships within datasets to detect corruption due to EMI.

Description: The notion of data may include individual data values and also collections of data items, such as lists, arrays, records and sets. The credibility checking can include range checking, and consistency of values between related data, such as by using the technique known as 'median filtering'.

There are several aspects to consider such as static (compile time) data typing, static range checking and dynamic (at runtime) value range checking, both on assignment and during the evaluation of arithmetic expressions.

Identification: Various checking schemes can be used to enable detection of corruption, for example, checksums or CRCs.

Various techniques described in this section can be used at the hardware level to implement an acceptable solution.

Mitigation: Upon detection of an anomaly, apply an appropriate response as defined in the safety documentation.

For a safety-related system, transition to a safe state as recorded in the safety documentation.

For a sub-system or element, transition to a DS as recorded in the safety manual, which shall provide sufficient detail on it to allow its correct use by a safety system's designer.

However, by using redundant electromagnetically-diverse-technology channels (see Section 2.2.3) it may be practicable to continue safe operation by switching from a failed channel to one that is still operating correctly.

Reference: [139]

Note: The importance of this technique depends on whether the safety function is intended for continuous operation or on demand.

2.3.19 Error detection/correction by monitoring program sequence (i.e. watchdogs)

Global aim: To detect a defective program sequence or timing and either take appropriate actions to maintain the safety integrity level/systematic capability; or restart the correct sequence if this is appropriate for maintaining the safety integrity level/systematic capability.

A defective program sequence exists if the individual elements of a program (for example software modules, subprograms or commands) are processed in the wrong sequence or period of time, or if the clock of the processor is faulty.

Global mitigation: Upon detection, apply an appropriate response as defined in the safety documentation.

For a safety-related system, transition to a safe state as recorded in the safety documentation.

For a sub-system or element, transition to a DS as recorded in the safety manual, which shall provide sufficient detail on it to allow its correct use by a safety system's designer.

However, by using redundant electromagnetically-diverse-technology channels (see Section 2.2.3) it may be practicable to continue safe operation by switching from a failed channel to one that is still operating correctly.

2.3.19.1 Watchdog with separate time base without time-window

Description: External timing elements with a separate time base (for example, watchdog timers) are periodically triggered to monitor the computer's behaviour and the plausibility of the program sequence. It is important that there is a clear design justification for the placement of triggering points in the program.

The watchdog is not triggered at a fixed period, but a maximum interval is specified.

The watchdog(s) should be designed using appropriate electromagnetic resilience techniques and measures that comply with this Code of Practice.

Identification: When the program fails to trigger any watchdog, a failure is indicated.

There could be several watchdogs, each monitoring different points in the program's execution sequence.

2.3.19.2 Watchdog with separate time base and time-window

Description: Timing elements physically separate from the computer, with a separate time base (watchdog timers), are periodically triggered to monitor the computer's behaviour and the plausibility of the program sequence.

It is important that there is a clear design justification for the placement of triggering points in the program.

Lower and upper time limits are given for the watchdog.

This technique is preferred over Section 2.3.19.1.

Identification: If the program sequence takes a longer or shorter time than expected, a failure is indicated.

2.3.19.3 Logical monitoring of program sequence

Description: The correct sequence of the individual program sections is monitored using software (for example, counting procedure, key procedure) or using external monitoring facilities.

It is important that there is a clear design justification for the placement of triggering points in the program.

This technique is preferred over Section 2.3.19.1 above.

Identification: If the correct program sequence does not occur, a failure is indicated.

2.3.19.4 Combination of temporal and logical monitoring of program sequences

Description: A temporal facility (such as a watchdog timer with a time-window) monitoring the program sequence is retriggered only if the sequence of the program sections is also executed correctly.

This technique is preferred over any of the three techniques in Sections 2.3.19.1, 2.3.19.2 or 2.3.19.3 above. It is also preferred over the application of both Sections 2.3.19.2 and 2.3.19.3 at the same time but independently.

Identification: If the temporal facility monitoring the program sequence is not retriggered as required, a failure is indicated.

2.3.20 Error detection and correction using multi-channel input/output interfaces

Aim: To detect random hardware failures (stuck-at failures), failures caused by external influences (such as EMI), timing failures, addressing failures, drift failures and transient failures (such as intermittency).

Description: This is a dataflow-dependent multiple-channel technique with independent inputs and/or outputs for the detection of random hardware failures and systematic errors.

Identification: Failure detection is carried out by comparing the signals with each other.

The comparator (the circuit used to compare channels and detect errors) is a weak point and so must be designed to have considerably greater electromagnetic resilience for this technique to be effective (for example, very frequent dynamic testing). The technology used and the reliability and resilience of the comparator must be justified in the safety documentation.

Mitigation: If a signal corruption is detected by the communicating partner(s), retransmission of the input or output data is requested. If the failure clears, continue operation as usual.

However, if during the time available the failure does not disappear, apply an appropriate response as defined in the safety documentation.

For a safety-related system, transition to a safe state as recorded in the safety documentation.

For a sub-system or element, transition to a DS as recorded in the safety manual, which shall provide sufficient detail on it to allow its correct use by a safety system's designer.

However, by using redundant electromagnetically-diverse-technology channels (see Section 2.2.3) it may be practicable to continue safe operation by switching from a failed channel to one that is still operating correctly.

Reference: [117]

2.3.21 Using test patterns: static and dynamic

Aim:

To detect static failures ('stuck-at' failures) and cross-talk, particularly in input and output units (digital, analogue, serial or parallel), and to prevent the sending of inadmissible inputs or outputs to the process.

Description:

This is a dataflow-independent cyclical test of input and output units.

It uses a defined test pattern to compare observations with the corresponding expected values.

The test pattern information, the test pattern reception, and test pattern evaluation must all be independent of each other.

The test pattern should not interfere with the correct operation of the safety function.

Useful for increasing electromagnetic resilience by detecting damage caused by over-voltages from lightning, electrostatic discharges or other sources.

Identification:

When the observations do not correspond with the expected values for the test pattern, a failure is indicated.

Mitigation:

Repeat the test pattern as many times as there is time for, without unacceptably degrading the safety integrity.

If during the time available the failure clears, log in the EDR (if one is available) and continue operation as usual.

If during the time available the failure does not clear, log in the EDR (if one is available) and apply an appropriate response as defined in the safety documentation.

For a safety-related system, transition to a safe state as recorded in the safety documentation.

For a sub-system or element, transition to a DS as recorded in the safety manual, which shall provide sufficient detail on it to allow its correct use by a safety system's designer.

However, by using redundant electromagnetically-diverse channels (see Section 2.2.3) it may be practicable to continue safe operation by switching from a failed channel to one that is still operating correctly.

2.3.22 Using fibre-optic cables for signals and data communications

Aim:

Avoid the effects of electromagnetic disturbances on communications media by using metal-free fibre-optic cables, which do not conduct electromagnetic disturbances.

Description:

Optical fibres *in themselves* (see the Note below) are unaffected by all electromagnetic disturbances (although they require protection from the thermal effects of lightning strikes, if exposed to them) and with suitable environmental protection can be used in all applications, including the most arduous.

Fibre-optic cables and their electronic interfaces (transmitters and receivers) are available in a wide range of types (and costs) to carry analogue signals from DC up to several GHz and data at up to hundreds of GBaud.

Where electrical power requirements are under 5 Watts, it is also practicable to carry AC or DC power over ordinary fibre-optics, converting the optical power into electrical power by using a photovoltaic cell instead of a signal/data receiver.

Mitigation: Optical fibre transmitters and receivers themselves are affected by electromagnetic disturbances, and so need to employ the good electromagnetic design techniques described in Section 2.3.26.

However, they are very small, making it much easier and less costly to achieve a given level of risk-reduction in communications than when using metal cables.

Metal-free optical fibres are a good choice for at least one of the channels when designing an electromagnetically diverse multi-channel redundant communication link (see Section 2.2.3).

Note: Certain types of optical cables use metal foils as moisture barriers, metal wires as drawstrings, and/or metal armour. These all conduct electromagnetic disturbances and can also worsen the thermal effects of lightning if the optical cable is struck by lightning. Consequently, this technique recommends the use of metal-free optical cables.

2.3.23 Techniques and measures for AC and DC power supplies/power converters

Global aim: To detect or tolerate failures caused by degradations or defects in any of the electrical power supplies.

Global description: *Degradations and defects in both DC and AC supplies:*

Under voltages, over-voltages, sags, swells, and interruptions lasting from less than one microsecond to many hours, days, even months in some cases.

AC ripple and noises with any frequency range and level.

Transients, 'spikes' and surges lasting from less than one microsecond to hundreds of milliseconds.

Degradations and defects in AC supplies only:

Waveform distortions, frequency perturbations and, in multi-phase supplies, phase and/or voltage imbalances, all with any amplitudes and lasting from less than one second to many hours, days, even months in some cases. Includes incorrect phase rotation.

2.3.23.1 Detecting degradations and defects

Identification: Various devices and circuit techniques are readily available for detecting any/all defects in AC or DC power supplies. For detecting excessive RF noise, see Section 2.3.23.3.

Mitigation: Upon detecting a degradation or defect in a power supply, apply an appropriate response as defined in the safety documentation.

For a safety-related system, transition to a safe state as recorded in the safety documentation.

For a sub-system or element, transition to a DS as recorded in the safety manual, which shall provide sufficient detail on it to allow its correct use by a safety system's designer.

May usefully be combined with mitigation in Section 2.3.23.2 and/or Section 2.3.23.4.

2.3.23.2 Power hold-up

Aim: To maintain the power supply for long enough during and/or after any transient or short-term deficiencies in the electrical power supply (such as dips, dropouts interruptions, under voltages, sags, etc.) to avoid a dangerous failure.

Description: Sufficient energy is stored in capacitors, supercapacitors, batteries, etc., to ensure the above aims are met.

In the case of long sags, under voltages or interruptions, the energy storage should be sufficient to continue correct (safe) operation whilst the EUC is put into a safe state or some other action taken to maintain the safety integrity as described in the safety documentation.

EUCs with high power requirements and/or requiring a long time to be put into a safe state despite lack of power for the safety-related system might use large battery banks (for example either directly or as part of a UPS) and/or rotating reserve power generators.

Identification: Analysis and testing of the worst possible combinations of circumstances, including a continuous low and/or distorted supply voltage, components tolerances and the effects of ageing, to ensure that the above aims are reliably met.

Mitigation: Improvement of the design, for example, by adding more energy storage.

Before the energy storage becomes exhausted to the point where errors, malfunctions or failures could possibly occur, appropriate action shall be taken to maintain the safety integrity level/systematic capability.

For a safety-related system, transition to a safe state as recorded in the safety documentation.

For a sub-system or element, transition to a DS as recorded in the safety manual, which shall provide sufficient detail on it to allow its correct use by a safety system's designer.

May usefully be combined with Section 2.3.23.4.

2.3.23.3 Detecting excessive radio frequency noise on power supplies

Aim: To detect the presence of excessive noise on power supplies, whether caused by failed/degraded decoupling capacitors, shielding, filtering, etc., or by EMI.

Description: Simple broadband radio frequency (RF) detectors can readily be created using ordinary circuit techniques (for example, a resistor, Schottky diode, capacitor, and operational amplifier) that will reliably detect frequencies up to tens of MHz. Some semiconductor manufacturers make single-chip RF detectors guaranteed to detect up to many GHz.

It will generally be necessary to set the sensitivity of the detector so that it does not trigger on the normal systematic noises made by the equipment or system itself in any operating mode when operating correctly.

Identification: Excessive levels of RF on AC power lines or DC power rails cause the RF detector to trigger.

Mitigation: Apply an appropriate response as defined in the safety documentation.

For a safety-related system, transition to a safe state as recorded in the safety documentation.

For a sub-system or element, transition to a DS as recorded in the safety manual, which shall provide sufficient detail on it to allow its correct use by a safety system's designer.

2.3.23.4 Redundant electromagnetically diverse power supplies

Aim: To maintain the required safety integrity level/systematic capability despite any of the problems detected by Sections 2.3.23.1, 2.3.23.2 and 2.3.23.3, by providing alternative power supplies.

Description: Providing alternative power supplies to replace failed ones.

Identification: Problems with power supplies may be detected using the techniques and measures described in Sections 2.3.23.1, 2.3.23.2 and 2.3.23.3 above.

Mitigation: The availability of redundant electromagnetically diverse power supplies (see Section 2.2.3) allows safe operation by switching from a failed power supply to one that is still operating correctly. The switch must be very electromagnetically robust as described in the safety documentation.

2.3.24 Monitoring of ventilation, cooling and heating

Aim: Failures of the ventilation, cooling or heating may expose the safety-related system to excessive environmental conditions, possibly increasing the rate of dangerous failure to an unacceptable level. Such failures could be caused by electromagnetic disturbances.

Description/ identification: Ventilation, cooling and heating systems are monitored for correct operation.

Mitigation: When a failure is detected, an appropriate response is made as defined in the safety documentation, before the EUC or its safety-related system is adversely affected.

For a safety-related system, transition to a safe state as recorded in the safety documentation.

For a sub-system or element, transition to a DS as recorded in the safety manual, which shall provide sufficient detail on it to allow its correct use by a safety system's designer.

However, by using redundant electromagnetically diverse ventilation, cooling or heating systems (see Section 2.2.3) it may be practicable to continue safe operation by switching from a failed one to one that is still operating correctly.

2.3.25 Careful use of wireless (radio) data communications

Aim: Where wireless communication is necessary for the system design, ensure that any wireless malfunction due to unwanted (in-band) and/ or co-channel interference will not cause an unsafe failure, and that the introduction of a wireless function does not adversely impact upon other safety related parts of the system.

Description: As many products now include an element of wireless functionality, it is conceivable that they will be required to contribute to the level of safety. A 'heart beat' signal is typically employed in wireless design to ensure that there is continuous communication between the transmitter and receiver.

Selection of suitable frequencies that support continuous transmission is required since many frequency allocations do not allow for this type of transmission. Reference [140] provides recommendations on suitable frequencies, power levels and modulation techniques for short-range wireless systems with implications for the safety of human life.

The introduction of a wireless function will change the electromagnetic environment and the compatibility of other safety-related parts of the system needs to be ensured by sufficient immunity at the frequencies of wireless operation, plus techniques and measures such as those in this Code of Practice, which ensure that even if the immunity is insufficient for any reason, safety integrity level/systematic capability is maintained.

Mitigation:
Where the 'heart beat' signal is lost, a defined signal is generated and fed into the system. If during the time available the heart beat signal is re-established, log this in the EDR (if one is available) and continue operation as usual.

If during the time available the 'heart beat' signal is not re-established, log in the EDR (if one is available) and apply an appropriate response as defined in the safety documentation.

For a safety-related system, transition to a safe state as recorded in the safety documentation.

For a sub-system or element, transition to a DS as recorded in the safety manual, which shall provide sufficient detail on it to allow its correct use by a safety system's designer.

However, by using electromagnetically-diverse redundant channels (see Section 2.2.3) it may be practicable to continue safe operation by switching from the failed wireless link to another data communication link that is still operating correctly.

2.3.26 Good electromagnetic engineering at every level of design

Aim:
To use accepted, good electromagnetic engineering practices at the time of system implementation so that a first line of defence against electromagnetic disturbances is provided.

Description:
Well-proven and widely-accepted good electromagnetic engineering design practices at the time of system implementation are applied at every level of design as appropriate, including (but not limited to) partitioning printed circuit boards (PCBs), units/modules/subassemblies/products, systems, installations, networks, etc. into different electromagnetic zones (see [24]), and also into lightning protection zones (usually the same as the electromagnetic zones) see [36], segregated by physical space and/or other electromagnetic mitigation techniques.

Identification:
Design assessment by persons competent in the relevant electromagnetic design issues.

Mitigation:
By competent correction of the design, where required.

References:
(a) For circuits, units, modules, subassemblies, products, etc., see Annex B.3.
(b) For cabinets, systems, installations, networks, etc., see Annex B.2.
(c) The 'Electromagnetic Zoning' technique [24] and guides based upon it: [21]-[23].

Examples include:

(a) electronic/electrical design appropriate for each electromagnetic zone;
(b) selection of electronic, electromechanical and electrical components appropriate for each electromagnetic zone;
(c) communications design (within and between electromagnetic zones);
(d) PCB design and layout (often incorporates several electromagnetic zones);
(e) power converter design e.g. AC-DC, DC-DC, DC-AC, AC-AC (generally located at electromagnetic zone boundaries);
(f) enclosure design for units/modules/subassemblies and products (could incorporate several electromagnetic zones);
(g) mitigation techniques such as filtering, shielding, galvanic isolation, surge and transient suppression, etc. (located at electromagnetic zone boundaries);
(h) system design (generally incorporates several electromagnetic zones); and
(i) installation and network design (always incorporating several electromagnetic zones).

2.3.27 Design to comply with EMC test specifications as set out in Sections 2.1.3 and 2.1.4

Aims: To help ensure that the safety-related system, or sub-systems or elements intended to be used in a safety-related system, will comply with the EMC test specifications as set out in Section 2.1.3 and, if relevant, Section 2.1.4, during verification and validation (see Section 2.5.2) if/when they are tested.

Identification: Achieved through regular assessment by personnel who are competent in the electromagnetic design of the relevant hardware and/or software, commensurate with the safety integrity level/systematic capability.

Mitigation: Modification of the design, followed by re-assessment, until the appointed assessors are satisfied.

Reference: See Clause 8 of IEC 61508 [8] for guidance on the degree of independence required for the assessors.

2.3.28 De-rating of hardware components, where appropriate

Aim: To increase the reliability of hardware components, particularly those used for the suppression of electromagnetic disturbances or protection against their effects.

Description: Hardware components are operated at levels well below their specified maximum ratings or stress levels.

EMI suppression/protection components shall be especially conservatively rated to survive repeated stress levels considerably higher than the worst anticipated, taking into account the full range of all reasonably foreseeable physical and climatic environments over the lifecycle (such as vibration, shock, humidity, extremes of ambient temperature (for example, when the air-conditioning has failed), etc.).

Identification: Achieved through independent assessment of the design by personnel who are competent (as required for the safety integrity level/systematic capability) in the field reliability of the hardware components concerned.

Mitigation: By modification of the design, followed by re-assessment, until the appointed assessors are satisfied.

References: See Clause 8 of IEC 61508 [8] for guidance on the degree of independence required for the assessors.

2.4 Techniques and measures for implementation, integration, installation and commissioning

2.4.1 Providing information on constraints and additional measures

Aims: To aid procurement, installation and commissioning in accordance with the relevant design requirements specification for electromagnetic resilience.

Identification: Achieved through the assessment of the intended operational site and its characteristics by personnel who are competent in the relevant site-related issues, commensurate with the safety integrity level/systematic capability.

Mitigation: By modification of the design, followed by re-assessment, until the appointed assessors are satisfied.

Measures required include, but are not limited to, the provision of information on:

(a) any constraints on the physical positioning of the items of equipment that comprise the safety-related system;

(b) any constraints on types, lengths and routing of power, control and signal interconnecting cables;

(c) the methods to be used when terminating any cable screens (shields);

(d) the types of connectors to be used and any special assembly requirements;

(e) the electrical power supply requirements (power quality);

(f) any additional screening (shielding) required, and how it should be installed;

(g) any additional filtering required, and how it should be installed;

(h) any additional overvoltage and/or overcurrent protection required, and how it should be installed (for example, by referencing the appropriate requirements in IEC 62305 series);

(i) any additional power conditioning required (such as a reliable UPS);

(j) any additional electrostatic discharge protection requirements (such as control of humidity);

(k) any additional physical protection required (for example, against the possibility of unusual physical and/or climatic conditions);

(l) the earthing (grounding) and bonding requirements for the installation;

(m) the procedures and materials to be used;

(n) any protection that is required against corrosion and its effects over the lifecycle; and

(o) any operator/maintainer constraints, for example, the use of mobile phones or cellphones whilst performing commissioning or maintenance.

In addition: proper installation and commissioning, having regard to the constraints and additional measures listed above (and any others not listed above), should be competently verified before the system is first operated (see Section 2.5.4) and thereafter checked regularly during its lifecycle, depending on the safety integrity level/systematic capability (see Section 2.6.2).

References: See Clause 8 of IEC 61508 [8] for guidance on the degree of independence required for the assessors.

2.4.2 Procuring materials, components and products

Aim: To ensure that all materials, components and products are procured according to their specifications for achieving electromagnetic resilience. Substandard or counterfeit materials, components and products are increasingly appearing in supply chains, especially when purchased on the 'grey market' (an activity that this Code of Practice does not recommend).

To help ensure that the safety-related system will comply with the EMC test specifications from Sections 2.1.3 and 2.1.4, during verification and validation (see Section 2.5.2).

Identification: By regular quality audits on goods received during the project.

Audits shall be carried out by personnel who are competent in the relevant quality control issues for the types of goods concerned in each case, commensurate with the safety integrity level/systematic capability.

Appropriate tests (see Section 2.5.2 and, if appropriate, Section 2.5.3) should be applied to verify suppliers' claims of compliance with specifications, the rate of which depends on the acceptable quality level (AQL) chosen in each case.

Such tests are recommended in general to avoid substandard or counterfeit materials, components and products from being incorporated in the safety-related system.

Mitigation: Replacement of the out-of-specification materials, components or products with in-specification materials, components or products that satisfy the appointed inspectors, before they are assembled.

References: See Clause 8 of IEC 61508 [8] for guidance on the degree of independence required for the assessors.

Example 1

In the military avionics industry, it is not unknown to hear claims that suppliers' build quality slips by enough to cause failure to meet specifications after seven units have been manufactured. Detecting the failure and ensuring that the supplier corrects the problem is claimed to typically result in a further failure to meet specification, another seven units later.

Example 2

The US Department of Defence has found counterfeit components in every weapons system, and in response has created a Regulation which all its suppliers are now required to comply with, to try to prevent counterfeits from entering the military supply chain [9].

2.4.3 Assemble/integrate according to the electromagnetic resilience design

Aims: To ensure that the correct materials, components and products are used in the correct ways so that the safety-related system and its sub-systems and elements are assembled and integrated according to their design requirements specifications for achieving electromagnetic resilience.

To ensure that good electromagnetic engineering practices are employed (see Section 2.3.26) as appropriate during assembly and integration.

To help ensure that the safety-related system will comply with the EMC test specifications as set out in Sections 2.1.3 and 2.1.4, during verification and validation (see Section 2.5.2).

Identification: By regular quality audits and/or assessments by personnel who are competent in the assembly/integration activities concerned, commensurate with the safety integrity level/systematic capability.

Mitigation: Replacement of incorrect materials, components or products, and/or reworking of incorrect assembly or integration, as required, to satisfy the appointed assessors.

References: See Clause 8 of IEC 61508 [8] for guidance on the degree of independence required for the assessors.

2.4.4 Install/commission according to the design for achieving electromagnetic resilience

Aims:
To ensure that the correct installation and commissioning methods are used for the safety-related system and its sub-systems and elements, according to their associated design requirements specifications for achieving electromagnetic resilience.

To ensure that good electromagnetic engineering practices are employed (see Section 2.3.25) as appropriate during installation and commissioning.

To help ensure that the safety-related system will comply with the EMC test specifications from Sections 2.1.3 and 2.1.4 during verification and validation (see Section 2.5.2).

Identification:
By regular quality audits and/or assessments by personnel who are competent in the installation/commissioning activities concerned, commensurate with the safety integrity level/systematic capability.

Mitigation:
Reworking of incorrect installation or commissioning as required to satisfy the appointed assessors.

References:
See Clause 8 of IEC 61508 [8] for guidance on the degree of independence required for the assessors.

2.5 Techniques and measures for verification and validation (including testing)

2.5.1 Applying verification and/or validation techniques and measures

Aims:
To verify and/or validate as far as is practicable that the design techniques and measures that have been applied function according to the relevant design requirements specification created as described in Section 2.1.

Identification:
By performing a sufficient number of techniques listed below, or equivalent techniques described in the safety documentation, to enable different types of weaknesses or omissions in the design to be discovered.

The competency, measurement accuracy and measurement uncertainty required for each verification or validation technique shall be commensurate with the safety integrity level/systematic capability.

Verification applies these techniques to all components, sub-assemblies, etc. of the safety-related system.

Where the component or sub-assembly is a third-party item, its manufacturer may have performed some or all of these techniques, and documented their results in the item's safety manual.

Validation applies these techniques at the highest practicable level of assembly of the safety-related system.

Failure prediction techniques may be helpful for quantitative risk assessment, when the risk cannot be shown to be tolerable through other qualitative means.

Typical quantitative techniques include:

(a) failure modes and effects analysis (FMEA);
(b) failure modes, effects and criticality analysis (FMECA);
(c) cause-consequence diagrams;
(d) event tree analysis (ETA);
(e) fault tree analysis (FTA); and
(f) fault tree models.

Mitigation: Changes are made to the design or operation to eliminate the weaknesses or omissions, and the relevant verification or validation re-applied.

Preceding lifecycle phases should be reviewed if they may be affected by the changes. Consideration should be given as to whether similar weaknesses or omissions may be present in other, similar safety functions. If so, similar changes should be carried out to those safety functions.

This process is repeated until the appointed assessors are satisfied. The decisions made and actions taken in this regard should be described in detail in the safety documentation.

Note 1: These quantitative techniques were not originally developed to deal with the effects of EMI, so they will need to be competently modified to take into account the issues mentioned in Section 2.1.2.

Note 2: Because there can be multiple orthogonal (i.e. independent) effects acting on equipment, Taguchi's 'Design of Experiments' approach can help improve tests for robustness by quickly determining the worst cases to be tested.

References: See the list in Annex B.15, especially [746].

Examples of verification and validation techniques:

(a) demonstrations, such as demonstrating that the functional safety requirements have been correctly implemented, using any appropriate methods.

(b) checklists, to ensure that design techniques and measures have been observed, applied and implemented correctly.

(c) inspections, to ensure that the designs for assembly and installation have been correctly followed.

(d) reviews and assessments, to ensure compliance with the objectives of each phase of the lifecycle. These are usually performed by competent persons on each phase of the lifecycle and the various stages of the activities within each phase.

(e) independent reviews and assessments. As for (d), with the degree of independence being related to the safety integrity level/systematic capability (see Clause 8 of Reference [8]).

(f) audits, which include verification processes for specification, design, assembly and installation.

(g) 'walk throughs' of normal operation and plausibly abnormal operations (sometimes called 'devil's advocacy').

(h) individual and/or integrated hardware tests. Different parts of the final assembly or system are assembled step by step, with checks and tests applied to ensure that they function correctly at each step.

(i) validated computer modelling, simulation, etc.

(j) the normal EMC tests applied in accordance with Section 2.5.3 can be modified to provide greater coverage of the possible effects of EMI, as described in [651] [601] and [602]; also see Section 2.5.4.

2.5.2 Verification testing to the EMC test plan from 2.1.3 and 2.1.4

Aims: To ensure that the safety-related system complies with the EMC test specifications from Section 2.1.3 and, if appropriate, Section 2.1.4.

Identification: By performing tests in accordance with the EMC test plan(s) created by applying Section 2.1.3 and (if appropriate) Section 2.1.4, using competent test personnel using calibrated test equipment and facilities.

Manufacturers are not necessarily precluded from doing these tests themselves or constrained from using certain types of third-party test laboratories.

Care shall be taken over the order of tests, for example, performing emissions after immunity to reveal whether seals or protection have been 'softened' during the immunity test.

The degree of accuracy, confidence, test accreditation and independence required for these tests is — like most functional safety issues — generally dependent on the safety integrity level/systematic capability.

Mitigation: Modification of the safety-related system, followed by re-verification and/or re-validation of the failed tests.

Depending on the tests which were failed, and the modifications required to achieve passes to them, it may be necessary to redo other EMC tests, possibly all of them.

This process is repeated until the appointed assessors are satisfied.

The decisions made and actions taken in this regard should be described in detail in the safety documentation.

Note: Complying with the conventional test standards alone is insufficient for EMI resilience (see Section 1 and References [11]-[14]).

2.5.3 Using non-standardised ad hoc checks or tests

Aims: To help ensure that the safety-related system or any component part of it has sufficient electromagnetic resilience, for the safety integrity level/systematic capability.

Complying with the EMC tests specified by Section 2.1.3 and (if appropriate) Section 2.1.4 is necessary for maintaining sufficiently high levels of EUC availability so that operators or owners are not inclined to modify or degrade the safety-related systems to achieve their productivity targets.

Of course this is very important for electromagnetic resilience, but no affordable or practicable EMC test plan can possibly demonstrate that electromagnetic disturbances cannot degrade the safety integrity level/systematic capability, even at level 1.

Employing a suitable number of design techniques and measures is what makes it possible for a safety integrity level/systematic capability not to be degraded by electromagnetic disturbances over the lifecycle. Section 2.5.12 lists techniques and measures that may be used to verify or validate this.

Non-standardised or ad hoc checks or tests can be used in addition to the list in Section 2.5.3 to achieve the necessary confidence in the electromagnetic resilience design, according to the safety integrity level/systematic capability.

In many situations they can prove very useful in assessing electromagnetic resilience design.

Identification: By performing non-standardised or ad hoc checks or tests.

Non-standard ad hoc test methods shall be justified by recording the following in the safety documentation:

(a) Rationale.

What needs to be measured? What is the purpose of measuring it? Why is a non-standard ad hoc test being proposed rather than a standards-based method?

(b) A detailed explanation of the test method.

Including figures or photographs and its theoretical underpinning.

(c) A demonstration of validity of this non-standard ad hoc test.

(If it is not immediately obvious to an engineer competent in testing the issue concerned).

Examples of some non-standard ad hoc checks and tests include:

(a) significantly increasing the test levels of standard immunity tests;
(b) modulating continuous wave (CW) disturbances with frequencies, pulse shapes or patterns, or wave shapes to which a design might be especially susceptible (from inspection/investigation of the design);
(c) applying two or more disturbances at once (for example, multiple frequencies during conducted or radiated tests to cause intermodulation in the tested design);
(d) applying different wave shapes on transient tests (such as surge, ESD, etc.);
(e) performing significantly larger numbers of transient tests to cover a greater proportion of the range of possible equipment states;
(f) checks on earthing, grounding and bonding by, for example, measurement with appropriate DC meters and/or visual inspections;
(g) checking that temperature and humidity sensors are functioning correctly (to help prevent corrosion of shielding, overheating of filters, etc.);
(h) checking the behaviour of shielding joints and gaskets during physical stress (for example, non-flat mounting surface), mechanical shocks, vibration, temperature changes, temperature extremes, condensation, icing, changes in air pressure (or water pressure for underwater equipment), etc., for example, by using battery-powered 'comparison noise emitters' inside an enclosure, and close-field probes outside it, within an environmental test chamber; and
(i) quick checks of emissions and immunity performance for units that have undergone highly-accelerated simulations of their lifecycle exposure to mechanical, climatic, chemical, etc., environments and/or user interactions (for example, opening/closing doors, hatches, inspection panels, etc.).

Mitigation: Modification of the safety-related system, followed by re-verification and/or re-validation of the failed checks or tests.

Depending on which checks or tests were failed, and the modifications required to achieve passes to them, it may be necessary to redo other checks or tests, or even standards-based testing.

This process is repeated until the appointed assessors are satisfied. The decisions made and actions taken in this regard should be described in detail in the project documentation.

References: [600]-[604] in Annex B.13.

A manufacturer is not precluded from doing these tests personally, or constrained to use certain types of third party test laboratories.

The degree of accuracy, confidence, and independence required for these non-standard ad hoc checks and tests is — like most functional safety issues — generally dependent on the safety integrity level/systematic capability.

2.5.4 Verifying correct installation and commissioning

Aims:

To ensure proper installation and commissioning having regard to the constraints and additional measures listed as the result of applying Sections 2.4.1 and 2.4.2, and any others not listed in those sections.

This is essential for 'version control' of the finished as-built system, listing all the hardware and software parts that have to be used together as a 'working set' to fulfil the initial functional safety objectives during the anticipated lifecycle.

Identification:

Inspection by competent personnel before the system is first operated and checked regularly during its lifecycle.

Mitigation:

Modification of the safety-related system, followed by re-inspection, repeated until the appointed assessors are satisfied.

The decisions made and actions taken in this regard should be described in detail in the safety documentation.

2.6 Techniques and measures for operation, maintenance, repair, refurbishment and upgrade

2.6.1 Assessment of changes in the electromagnetic environment

Aims:

To discover new electromagnetic environment conditions that were not taken into account in the original design.

To modify/upgrade as required so that availability is maintained at a high level (as discussed in Section 2.1.3).

Identification:

This is achieved by analysing the following, at least:

(a) changes in the EMC test standards used in the specification (see Sections 2.1.3 and 2.1.4);
(b) changes in the standards listed in Annex B.14;
(c) results from independent detection of electromagnetic disturbances as described in Section 2.2.9;
(d) results recorded as described in Section 2.2.5, and then analysed;
(e) the assessment techniques described in Section 2.1.3 and [650]; and
(f) proposed changes in the EUC, the safety-related system, or other equipment/systems that could affect inter-system or intra-system electromagnetic energy couplings into the safety-related system.

These proposed changes could be upgrades, repairs, or any modifications for any reasons.

It may be useful to perform a 'gap analysis', comparing the electromagnetic environment that was the basis of the original design requirements specification (see Section 2.1), with the current electromagnetic environment.

Mitigation: By re-applying the appropriate parts of the process described in this Code of Practice, as appropriate to the changes in the electromagnetic environment, in accordance with the approach taken by IEC 61508 in such circumstances.

This process is repeated until the appointed assessors are satisfied.

The decisions made and actions taken in this regard should be described in detail in the safety documentation.

2.6.2 Assessment of continuing correct installation

Aims: To ensure the maintenance of proper installation and commissioning having regard to the constraints and additional measures listed as the result of applying Section 2.4.1, and any others not listed in that section.

This activity is another essential for the 'version control' of the as-built system (see Section 2.5.4).

Identification: Regular inspections by competent personnel during the lifecycle.

These inspections may include, for example: grounding/bonding; shielding effectiveness; filter insertion loss; the condition of surge protection components/devices and electromagnetic shielding gaskets; unapproved modifications (including cable/connector replacements and/or additions, software upgrades or other changes); etc.

Mitigation: Modification of the safety-related system, followed by re-inspection, until the correct constraints and additional measures are once again satisfied.

Preventative maintenance shall also be employed wherever any aspect of the installation appears to be suffering from degradation of its electromagnetic characteristics at such a rate that they could become unacceptable before the next planned inspection.

Where the periodicity of the planned inspections is found to be inadequate to prevent certain electromagnetic characteristics from degrading by too much, the planning should be changed to inspect at least those characteristics sufficiently often that their degradation is corrected before they have degraded to the point of unacceptability.

Examples

A common example is conductive gaskets used to seal apertures in shielding enclosures, and dissimilar metal bonds (such as earth/ground connections). These are often subject to corrosion that progresses over time until they no longer function as well as required for electromagnetic resilience.

Another example is surge protection components and/or devices, which generally degrade as time progresses due to the surges they experience, until they can no longer provide adequate protection.

Note that these surge protection components/devices and earth/ground bonds might be located remotely from the electronics of the EUC and its safety system — for example, they may be installed as part of a site or vehicle's lightning and/or EMP protection system, and yet the electromagnetic resilience of the safety related system and/or its sub-systems or elements might still rely on the protection they provide.

2.6.3 Maintaining electromagnetic resilience despite modifications or changes

Aims:

To ensure that repairs, modifications, upgrades, refurbishment, etc. do not unacceptably degrade the electromagnetic resilience of the safety-related system.

For example, even replacing a cable with a different one, even if supposedly of the same type, can cause unacceptable degradation of electromagnetic emissions and/or immunity.

However, such issues should have been foreseen and taken care of in the planning (see Section 2.1.2), so that even if the EUC becomes unavailable as a result, it does not become unsafe.

This activity is another essential for the 'version control' of the as-built system (see Section 2.5.4).

Identification:

By re-applying the parts of the process described in this Code of Practice that are appropriate to the proposed changes to the safety-related system (this is in accordance with the approach taken by IEC 61508 in such circumstances).

Mitigation:

Implement whatever the above process shows to be necessary — whether in specifications, system design, detailed techniques and measures, verification/validation, etc. — to ensure that the repairs, modifications, upgrades, refurbishment, etc. do not unacceptably degrade the electromagnetic resilience of the safety-related system.

2.7 Electromagnetic resilience during decommissioning

Aims:

Functional safety, as defined in IEC 61508, covers the entire lifecycle including decommissioning (see Figure 1.3), so it is necessary to ensure that dismantling and/or disposal does not cause unacceptable functional safety risks due to the electromagnetic resilience of the relevant safety functions when a safety-related system is being degraded by the dismantling and/or disposal process even while those functions are still required.

> **Example**
>
> Certain types of batteries need controlled rates of charge and discharge if they are not to overheat and rupture, which would cause various kinds of safety hazards. In smaller batteries (such as laptop computers) these safety-related systems are built-into the battery, but for example in an electric traction vehicle they may be external items.
>
> Dismantling of such a vehicle might require that the charge/discharge safety-related system remains in full working order at all times for functional safety reasons, right up to the point of final disposal.
>
> Where it is not practicable to remove all power supplies (of any type: electrical, pneumatic, hydraulic, etc.) from an EUC, or when the EUC itself contains significant amounts of stored energy (electrical, pneumatic, hydraulic, nuclear fissionable or explosive materials, etc.), the relevant safety functions of its safety-related systems may need to be maintained in full operation until safe disposal has been achieved.

Identification: By re-applying the appropriate parts of the process described in this Code of Practice to the proposed dismantling and/or disposal project. (This is in accordance with the approach taken by IEC 61508 in such circumstances.)

Mitigation: Implement whatever the above process shows to be necessary — whether in specifications, system design, detailed techniques and measures, verification/validation, etc. — to ensure that the dismantling and/or disposal project does not unacceptably degrade the electromagnetic resilience of each safety-related system concerned.

Note: Dismantling and disposal may mean that the exposure of workers and/or the public to the foreseeable functional safety hazards is different from the operational stage of its lifecycle, and this may affect its safety integrity level/systematic capability. Its safety integrity level/systematic capability for the decommissioning stage might be higher or lower than during operation, for example.

Example

When a nuclear submarine or electronically-fused munition is removed in its entirety to a breaker's yard that has special protection measures and is far away from both military personnel and the public.

A changed safety integrity level/systematic capability will (of course) influence the process described in this Code of Practice.

2.8 Integrating third-party items into safety-related systems

2.8.1 The general iterative approach

Figure 2.1 shows an example of the iterative process by which volume-manufactured commercially-available standard products, used as elements of a safety-related system, are chosen based upon the electromagnetic resilience specifications of the safety-related system and/or its sub-systems or elements (see Section 2.1).

As Figure 2.1 shows, it may be necessary for the designer(s) to iterate the design of the electromagnetic mitigation measures, or even add new electromagnetic zones to create suitable electromagnetic environments for the chosen standard products.

In practice, this means that the detailed design of the safety-related system's electromagnetic resilience might have to be modified to achieve the specified safety integrity due to the characteristics of the chosen elements.

It should always be remembered that designing and realising any safety-related system is usually not a linear progression of steps - iteration (looping back to an earlier project stage) is often required as the real characteristics become apparent during the design, integration, implementation, installation, verification and validation stages.

Figure 2.1 shows:

Step 1: The electromagnetic resilience specifications for the safety-related system are developed using the techniques and measures in Section 2.1, comprising the list of EMC tests to be complied with plus a non-exhaustive list of appropriate electromagnetic resilience techniques and measures to be used.

Step 2: The electromagnetic resilience specifications for each of the sub-systems or elements to be used in the safety-related system are then developed from the electromagnetic resilience specifications created in Step 1. These comprise a specification for the EMC tests to be complied with, plus a list of electromagnetic resilience techniques and measures, taking into account any electromagnetic mitigation provided by the electromagnetic zone in which the sub-system or element will be located.

Step 3: The electromagnetic resilience specification for an individual sub-system or element is compared with the information provided by commercial suppliers in their products' safety manuals. These safety manuals should include details of the EMC tests that were complied with, the electromagnetic resilience techniques and measures employed, and (where continual error-free performance is not ensured) any DSs.

Step 4: The standard volume-manufactured sub-systems and elements to be incorporated into the safety-related system are chosen from the list of commercial products whose safety manuals meet the required specifications and have acceptable DSs.

Note that the required electromagnetic specifications should be included, in full detail, in the purchasing contract.

Also note that no reliance should be placed on CE marking or manufacturers' certificates/declarations of conformity.

Step 5: Where suitable commercially available products do not comply with the required specifications, electromagnetic mitigation measures and/or electromagnetic resilience techniques and measures may be applied, or existing mitigation, techniques or measures modified, at any level of assembly and in any electromagnetic zone in order to change the electromagnetic resilience specification for the individual sub-system or element in Step 2.

Step 6: Steps 2-4 are iterated for each sub-system and element until compliance is achieved with the electromagnetic resilience specifications for the safety-related system in Step 1.

Step 7: The same process is repeated for every sub-system or element in the safety-related system.

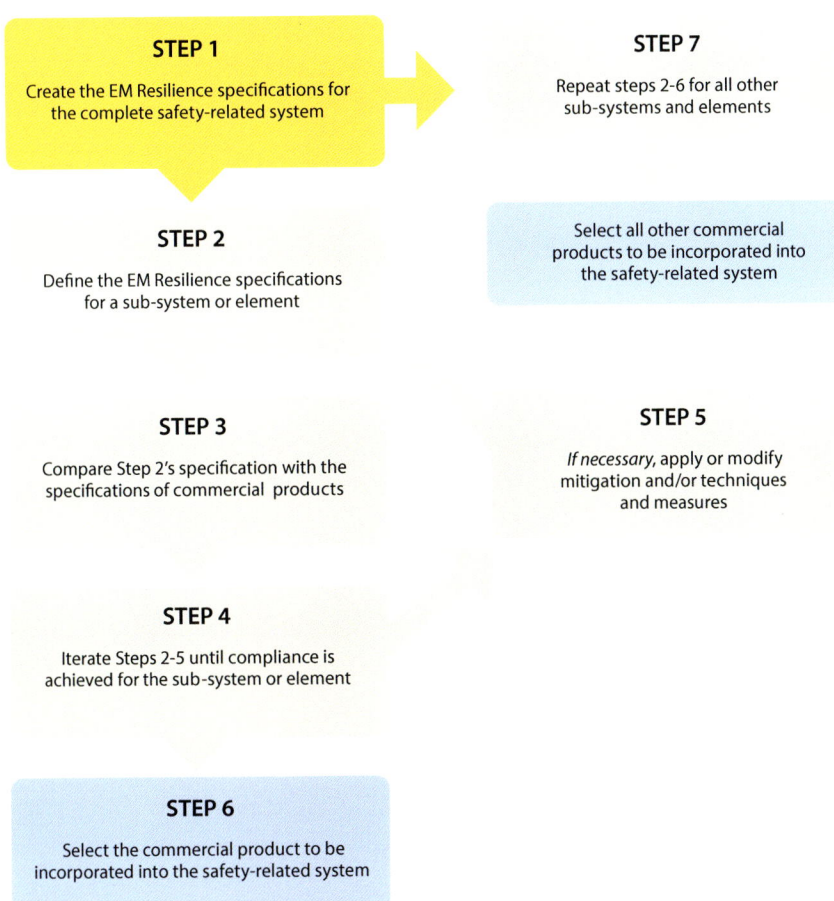

▼ **Figure 2.1** Choosing standard volume-manufactured sub-systems and elements for a safety-related system

STEP 1
Create the EM Resilience specifications for the complete safety-related system

STEP 2
Define the EM Resilience specifications for a sub-system or element

STEP 3
Compare Step 2's specification with the specifications of commercial products

STEP 4
Iterate Steps 2-5 until compliance is achieved for the sub-system or element

STEP 6
Select the commercial product to be incorporated into the safety-related system

STEP 7
Repeat steps 2-6 for all other sub-systems and elements

Select all other commercial products to be incorporated into the safety-related system

STEP 5
If necessary, apply or modify mitigation and/or techniques and measures

2.8.2 Suppliers' certifications and electromagnetic performance

Suppliers' markings, certifications, and declarations (including CE marking with respect to the EMC Directive) result from a 'self-declaration' process. Accordingly, it is recommended that companies involved with integrating any electronic equipment or system should take reasonable steps to check whether any markings, certificates or declarations issued by suppliers are reliably correct.

There are many independent assessment bodies that will validate and certify customer's products against their specifications. Using products whose EMC performance specifications are validated by independent assessment bodies is one way of achieving due diligence. Some suppliers are known to forge third-party assessment documents, so it is always a good idea to confirm them with the body purported to be the issuer.

Another way is to investigate suppliers' claims yourself, for example, by requesting test certificates or test reports; checking that they indicate the desired performance; and checking with the test laboratory to see how independent they are. Alternatively you could perform simple checks, or even full tests, to verify suppliers' performance claims.

The higher the safety integrity level/systematic capability to be achieved, the more work is required to achieve the assurance that purchased or free-issued elements have the electromagnetic characteristics their manufacturers claim.

2.8.3 Alternatively, use custom-manufactured elements

In some cases it could be a quite reasonable solution to pay a supplier of standard products to produce a custom-engineered version that meets the safety-system designers' electromagnetic specification and is provided with believable test results.

A product manufacturer might even be persuaded to make a completely new type of product for use as an element in a certain safety-related system.

This is typical of safety-related systems in automobiles (for example, anti-lock braking, engine management, etc.) where product volumes are high, justifying considerable investment in elements that can be incorporated by the safety-related system integrator without having to create special electromagnetic zones for the elements using electromagnetic mitigation.

The same 'custom-designed element' approach may also be appropriate where the safety-related systems are individually unique or only made in small quantities, for example, SIL 4 systems for nuclear power generating plant, nuclear fuel rod reprocessing, railway signalling and control, etc., which are not very price-sensitive.

Basic checklist for electromagnetic resilience techniques and measures

Note: *See Section 1.7 for information on the correct use and understanding of the 'importance' values and SIL levels (or their alternatives) listed in this Annex.*

Overview of some electromagnetic resilience techniques and measures	Importance				References in this Code of Practice	Applied: Yes/No Add any comments, references, and links
	SIL1	SIL2	SIL3	SIL4		
Techniques and measures for use in project management, planning and specification					2.1	
Project management and planning						
The processes for the management, planning, selection, design, implementation, commissioning, modification verification and maintenance of each safety function should explicitly include electromagnetic resilience measures and be documented.	M	M	M	M	2.1.1	
A competent person should have the overall responsibility for managing the electromagnetic resilience of the system. Appropriate competency should be made available at all lifecycle stages.						
Creating a design requirements specification						
To ensure that the design specification includes requirements for EMI, and that all reasonably foreseeable electromagnetic disturbances and their effects are taken into account in the specification of the system and its sub-systems and elements.						
Appropriate techniques and measures shall be defined and used to ensure that the safety-related system shall achieve the required SIL, and all of the sub-systems and elements incorporated within it shall achieve their required systematic capabilities, despite any electromagnetic disturbances over the lifecycle.	M	M	M	M	2.1.2	
Amongst other issues, the following shall be taken into account:						
(a) non-operation, when operation is required;						
(b) operation, when no operation is required; and						
(c) unintended or inaccurate operations						
The specification for electromagnetic resilience techniques and measures shall be (as far as is possible): complete, free from errors and contradictions, and easy to verify.						

Overview of some electromagnetic resilience techniques and measures	Importance				References in this Code of Practice	Applied: Yes/ No	Add any comments, references, and links
	SIL1	SIL2	SIL3	SIL4			
Specifying EMC test standards to help ensure the availability of the EUC							
To help ensure the availability of the EUC and its safety-related systems, throughout its lifecycle, so that safety-related systems continue to provide safe operation, taking into account availability, throughput rate, production rate, or other financial or mission-critical requirements.	M	M	M	M	2.1.3		
Protecting against high impact, unusual and malicious EMI							
To help achieve functional safety where high impact, unusual and malicious electromagnetic disturbances could reasonably foreseeably occur and cause temporary disturbance and/or permanent damage to hardware (electronic components, interconnections, etc.).	M	M	M	M	2.1.4		
Other technique or measure used in project management and planning:	-	-	-	-			
Other technique or measure used in project management and planning:	-	-	-	-			
Other technique or measure used in project management and planning:	-	-	-	-			
Techniques and measures for use in system design					2.2		
Separating safety-related system parts from non-safety-related system parts	HR	HR	HR	HR	2.2.1		
Recording how the design requirements are implemented through design choices	HR	HR	M	M	2.2.2		
Co-design of electromagnetically diverse hardware and software in multiple redundant channels	R	R	HR	HR	2.2.3		
To detect and/or correct systematic failures, using multiple electromagnetically diverse hardware channels and/or software components, to reduce the likelihood that the common-cause characteristics of electromagnetic disturbances will cause an incorrect output to be created.							
Hardware and software designers should work together (i.e. co-design) to achieve the required overall diversity in the most effective way in order to meet the requirements of the design requirements specification and its required safety integrity levels and/or systematic capabilities.							

Overview of some electromagnetic resilience techniques and measures

	Importance				References in this Code of Practice	Applied: Yes/No	Add any comments, references, and links
	SIL1	SIL2	SIL3	SIL4			
System integration, installation and commissioning							
To ensure that electromagnetic resilience is correctly considered when separately tested parts are brought together to form the complete functional system.	HR	HR	M	M	2.2.4		
Fault detection and event data recording for later diagnosis							
To increase the probability of localising malfunctions caused by electromagnetic disturbances.	R	R	HR	HR	2.2.5		
Improving the electromagnetic resilience of communication links, by using hardware and/or software techniques, to improve the reliability of the links							
Error detection							
Redundant data is appended to the actual data using error detection coding (EDC) techniques such as parity or cyclic redundancy checking (CRC) (see 2.3.11, 2.3.12, and 2.3.14), or suitable equivalent EDC techniques, to detect data corruption.							
Upon detection of data corruption, appropriate action is taken to maintain the safety integrity level/systematic capability, as described in the safety documentation. For example, various retry schemes could be used to improve the reliability of the link (at the expense of overall system performance).	HR	HR	HR	HR	2.2.6.1		
Where the safety manual for a sub-system or element includes a DS, it shall provide sufficient detail for correct use by a safety system designer.							
Error correction							
A variation of error detection using code such that a level of error correction is possible in order to both detect corruption and correct for its effects.							
Various error correcting code (ECC) schemes (see 2.3.11, 2.3.12, and 2.3.14) can be used to improve the reliability of the link at the expense of reduced data rate.	HR	HR	HR	HR	2.2.6.2		
Whenever error correction occurs, this should be logged to aid later diagnosis. See 2.2.5.							

Overview of some electromagnetic resilience techniques and measures

	Importance				References in this Code of Practice	Applied: Yes/No	Add any comments, references, and links
	SIL1	SIL2	SIL3	SIL4			
Protection of a sequence							
Extra sequence codes can be appended to each packet to enable detection of delayed, lost or duplicated packets.							
Various techniques and measures in this Code of Practice can be used at the packet level, e.g. just a single bit can be alternated between packets to detect a single packet failure (omission or duplication).	HR	HR	HR	HR	2.2.6.3		
More elaborate techniques can detect multiple packet failures or corruption.							
Synchronisation and re-synchronisation techniques — Synchronous system safety functions intended for continuous operation.	HR	HR	HR	HR			
Synchronous systems safety functions intended to operate on demand.	R	R	R	R	2.2.7		
Any kind of synchronous system that has no safe state.	M	M	M	M			
Protection from persistent interference by monitoring retry counts — Systems intended for continuous operation.	HR	HR	HR	HR			
On-demand systems.	R	R	R	R	2.2.8		
Independent detection of electromagnetic disturbances and/or EMI	R	R	R	R	2.2.9		
Protection of systems from tampering via communication links to external systems — To conserve the safety integrity level/systematic capability of systems, sub-systems or elements that have external communication links, especially with the Internet, at least for electromagnetic resilience.	R	R	R	HR	2.2.10		

Overview of some electromagnetic resilience techniques and measures

Technique and measures	Importance				References in this Code of Practice	Applied: Yes/No Add any comments, references, and links
	SIL1	SIL2	SIL3	SIL4		
Robust, high-specification electromagnetic mitigation. Especially useful when degradation or interruption of functionality is not desired.	R	R	R	R	2.2.11	
Other technique or measure used in system design:	-	-	-	-		
Other technique or measure used in system design:	-	-	-	-		
Other technique or measure used in system design:	-	-	-	-		
Other technique or measure used in system design:	-	-	-	-		
Other technique or measure used in system design:	-	-	-	-		
Techniques and measures for use in operational design In these subsections below, techniques and measures are classified as either hardware or software based, but some techniques and measures may have equivalent representations in either hardware or software, which might be more effective in some useful manner.					2.3	
Developing appropriate operation and maintenance instructions For procedures that help to avoid EMI-induced failures during the operation and maintenance of a safety-related system or a sub-system or element used within a safety-related system.	HR	HR	HR	HR	2.3.1	
Designing appropriate maintenance techniques To make it practical to monitor the condition/performance of, and replace if necessary, electromagnetic mitigation items such as filters, surge suppressors, conductive gaskets, etc., which may have a limited operational life.	HR	HR	HR	HR	2.3.2	
Limiting the possibilities for operation and hence for mis-operation. To help avoid EMI causing failures by affecting operator controls.	HR	HR	HR	HR	2.3.3	
Protecting against operator errors, mistakes and other foreseeable misuse	HR	HR	HR	HR	2.3.4	

Overview of some electromagnetic resilience techniques and measures

	Importance SIL1	SIL2	SIL3	SIL4	References in this Code of Practice	Applied: Yes/No Add any comments, references, and links
Protecting against hardware/software modifications or manipulations						
Using any technical means	HR	HR	HR	HR	2.3.5	
Range checking in hardware and in software						
Defensive programming techniques — To design software programs to detect anomalous control flow, data flow, or data values, which may have been caused by EMI during their execution, and to react in a predetermined and acceptable manner. Range checking the values of all variables (not just I/Os), sometimes called strong data typing. A number of bands are defined for the value of each variable. A typical 3-band example is: **(a)** normal operation; **(b)** warning zone; and **(c)** out of range.	R	R	HR	HR	2.3.6.1	
Sequence checking — Safety functions intended for continuous operation	HR	HR	HR	HR	2.3.6.2	
Safety functions intended for on-demand operation	R	R	R	R		
Correct rounding and resolution in all calculations	HR	HR	HR	HR	2.3.6.3	
Limited use of interrupts — To help reduce the impact of corruption due to EMI upon program execution	R	HR	HR	M	2.3.7	
Limited use of memory address pointer variables to reduce impact of memory corruption — System-safety functions intended for continuous operation	HR	HR	HR	HR	2.3.8	
System-safety functions intended for on-demand operation	R	R	R	R		
Any/all systems with no safe state	M	M	M	M		

Overview of some electromagnetic resilience techniques and measures

Overview of some electromagnetic resilience techniques and measures	Importance				References in this Code of Practice	Applied: Yes/No Add any comments, references, and links
	SIL1	SIL2	SIL3	SIL4		
Avoiding recursion To help reduce the impact of corruption due to EMI on program execution	HR	HR	HR	HR	2.3.9	
Error detection and correction for invariable memory (i.e. ROM or program memory) — **Signature of a word or block of data** To detect single and multi-bit corruption within a block of data. Various checking techniques are available, such as: Cyclic Redundancy Checks (CRC), Secure Hash Algorithm (SHA),and Hamming Codes (for correction as well as detection).	R	R	HR	HR	2.3.10.1	
Block replication with inversion to detect all bit failures	HR	HR	HR	HR	2.3.10.2	
Plus the use of electromagnetically diverse memories to improve effectiveness.	R	R	HR	HR	2.2.3	
Memory boundary protection To prevent incorrect areas being overwritten in the following types of memory: program; stack; statically-allocated variables; heap (dynamically allocated variables); inputs, and outputs	R	R	HR	HR	2.3.10.3	
Error detection and error correction techniques in redundant designs System-safety functions intended for continuous operation	HR	HR	HR	HR	2.3.11	
System-safety functions intended for on-demand operation	R	R	HR	HR	2.3.11	
Time-based error detection/correction in buses and interfaces to detect transient failures	R	R	HR	HR	2.3.12	
Combine with error checking codes to protect the sequence numbers or time codes.	R	R	HR	HR	2.3.11 & 2.3.13	

Overview of some electromagnetic resilience techniques and measures

	SIL1	SIL2	SIL3	SIL4	References in this Code of Practice	Applied: Yes/No Add any comments, references, and links
Error detection and error correction for variable memory (e.g. RAM). Detecting failures during addressing, writing, storing and reading data in memory.						
Memory testing — Before and/or during operation to detect memory-system-specific errors.	R	R	HR	HR	2.3.13.1	
One-bit redundancy — To detect some changes in the content of a memory location, bus or I/O register.	R	R	R	R	2.3.13.2	
Block replication with inversion to detect all bit failures — Using diverse types of memory can improve the effectiveness of this technique.	HR	HR	HR	HR	2.3.13.3	
	R	R	HR	HR	3.3.3	
Memory boundary protection	R	R	HR	HR	2.3.13.4	
Error detection/correction in ROM, RAM, buses and interfaces — Detects/corrects one or more bit failures in a word.	R	R	HR	HR	2.3.14	
Self-test supported by hardware (one-channel)	HR	HR	HR	HR	2.3.15.1	
Coded processing (one-channel)	R	R	R	R	2.3.15.2	
Error detection for logic and data processing units						
Reciprocal comparison by software — Two or more electromagnetically diverse processing units exchange data (results, intermediate results, and test data) and cross-check at defined 'restore points' from which system operation could be continued in the event of a fault.	HR	HR	HR	HR	2.3.15.3	
Self-test by software during operation	R	R	HR	HR	2.3.15.4	

Overview of some electromagnetic resilience techniques and measures		Importance				References in this Code of Practice	Applied: Yes/No Add any comments, references, and links
		SIL1	SIL2	SIL3	SIL4		
Error detection and error correction for electrical and electromechanical components To help control failures in components such as relays, actuators, magnetic logic devices etc. When using redundancy, diverse hardware and/or software improves the effectiveness as regards the common-cause effects typical of electromagnetic disturbances.		HR	HR	HR	HR	2.3.16 2.2.3	
Caution when using hardware or software libraries To ensure that the electromagnetic resilience measures are applied and verified throughout the whole software design and implementation of a safety related system.		HR	HR	M	M	2.3.17	
Error detection and correction for electronic components	**Testing by redundant hardware** To monitor the operation of the relevant function.	R	R	R	R	2.3.18.1	
	Using dynamic signalling techniques To detect static failures in communications and processing.	R	R	R	R	2.3.18.2	
	Caution with use of test access ports and boundary-scan To prevent any tests/diagnostics from making the system more susceptible to electromagnetic disturbances.	R	R	R	R	2.3.18.3	
	Monitored redundancy Compares the behaviour of two or more electromagnetically diverse (see 2.2.3) channels.	R	R	HR	HR	2.3.18.4	
	Hardware with automatic self-test To detect faults by periodic checking of the safety functions using automatic self-tests.	R	R	R	R	2.3.18.5	

Overview of some electromagnetic resilience techniques and measures

			Importance				References in this Code of Practice	Applied: Yes/No	Add any comments, references, and links
			SIL1	SIL2	SIL3	SIL4			
	Analogue signal monitoring To improve confidence in signals and controls.		HR	HR	HR	HR	2.3.18.6		
	'Data assurance' (content credibility checking) Uses known relationships within a dataset to detect corruption due to EMI.	System-safety functions for continuous operation.	HR	HR	HR	HR	2.3.18.7		
		System-safety functions for on-demand operation.	R	R	R	R			
	Watchdog (temporal monitoring) with separate time base without time-window *– only to be used if 2.3.19.2 or 2.3.19.3 below cannot be used.*		R	R	NR	NR	2.3.19.1		
Error detection and correction by monitoring program sequence (i.e. 'watchdogs')	**Watchdog (temporal monitoring) with separate time base and time-window** Periodically triggered to monitor the computer's behaviour and the plausibility of the program sequence, with both lower and upper time limits set; preferred over 2.3.19.1.		HR	HR	HR	HR	2.3.19.2		
	Logical monitoring of program sequence Monitoring of individual program sections using software (e.g. counting procedure, key procedure) or using external monitoring facilities; preferred over 2.3.19.1.		R	R	HR	HR	2.3.19.3		

Overview of some electromagnetic resilience techniques and measures

Technique/measure	Importance SIL1	SIL2	SIL3	SIL4	References in this Code of Practice	Applied: Yes/No	Add any comments, references, and links
Combination of temporal and logical monitoring of program sequences Combining both temporal (with time window) and logical monitoring to retrigger a temporal facility (e.g. an external watchdog) only if the sequence of the program sections is executed correctly. Preferred over either 2.3.19.2 or 2.3.19.3 above. Also preferred over 2.3.19.1 and 2.3.19.1 used together but independently.	R	R	HR	HR	2.3.19.4		
Error detection and error correction by comparing multi-channel input/output interfaces					2.3.20		
Using electromagnetically diverse hardware and/or software To improve the effectiveness of this technique as regards the common-cause effects typical of electromagnetic disturbances, permitting more confident error correction.	R	R	HR	HR	2.2.3		
Using test patterns: static and dynamic Using static and dynamic test patterns to detect static failures ('stuck-at' failures) and cross-talk, particularly in input and output units (digital, analogue, serial or parallel), and to prevent the sending of inadmissible inputs or outputs to the process.					2.3.21		
Using electromagnetically-diverse channels To permit more confident error correction.	HR	HR	HR	HR	2.2.3		
Use metal-free fibre-optic cables for signals and data They are intrinsically immune to electromagnetic disturbances.	R	R	R	HR	2.3.22		

Overview of some electromagnetic resilience techniques and measures

Techniques for AC and DC power supplies and power converters

To detect or tolerate failures caused by degradations or defects in any of the electrical power supplies.

	Importance				References in this Code of Practice	Applied: Yes/No	Add any comments, references, and links
	SIL1	SIL2	SIL3	SIL4			
Detecting degradations and defects Various devices and circuit techniques are readily available for detecting any/all defects in AC or DC power supplies.	HR	HR	HR	HR	2.3.23.1 3.3.3		
Power hold-up Using sufficient energy storage (e.g. batteries, supercapacitors, etc.) or back-up power supplies (e.g. generators) with appropriate action taken to maintain the safety integrity/systematic capability when the energy storage runs out.	HR	HR	HR	HR	2.3.23.2		
Detecting excessive radio frequency noise on power supplies	R	R	R	HR	2.3.23.3		
Redundant electromagnetically diverse power supplies Using redundant electromagnetically-diverse power supplies to continue safe operation by switching from a failed power supply to one that is still operating correctly (e.g. a backup/reserve power supply).	R	R	HR	HR	2.3.23.4		
Monitoring of ventilation, cooling and heating To detect whether they have been influenced by electromagnetic disturbances.	R	R	HR	HR	2.3.24		

Overview of some electromagnetic resilience techniques and measures	Importance				References in this Code of Practice	Applied: Yes/No Add any comments, references, and links
	SIL1	SIL2	SIL3	SIL4		
Careful use of wireless (radio) data communications Ensuring that wireless (radio) data communications will not cause an unsafe failure, and will not adversely impact other safety-related parts of the system.	HR	HR	M	M	2.3.25	
Good electromagnetic engineering practices used at every level of design To use accepted, good electromagnetic engineering practices at the time of system implementation in order to provide a first line of defence against electromagnetic disturbances.	HR	HR	HR	HR	2.3.26	
Design to comply with EMC test specifications from 2.1.3 (and 2.1.4 if appropriate) To help ensure that the safety-related system will comply with these EMC test specifications during verification and validation.	M	M	M	M	2.3.27	
De-rating of hardware components, where appropriate To increase the reliability of hardware components, particularly those used for the suppression of electromagnetic disturbances or protection against their effects.	R	R	R	HR	2.3.28	
Other technique or measure used in operational design:	-	-	-	-		
Other technique or measure used in operational design:	-	-	-	-		
Other technique or measure used in operational design:	-	-	-	-		
Techniques and measures for implementation, integration, installation and commissioning					2.4	

Overview of some electromagnetic resilience techniques and measures	Importance				References in this Code of Practice	Applied: Yes/No Add any comments, references, and links
	SIL1	SIL2	SIL3	SIL4		
Providing information on any constraints and/or additional measures required for installation and commissioning						
To aid installation and commissioning in accordance with the relevant design requirements for electromagnetic resilience.	HR	HR	HR	HR	2.4.1	
Procure materials, components and products						
According to their design specifications for achieving electromagnetic resilience.	HR	HR	M	M	2.4.2	
Assemble/integrate according to the electromagnetic resilience design						
Using the correct materials, components and products according to their design specifications for achieving electromagnetic resilience.	HR	HR	M	M	2.4.3	
Install/commission according to the design for achieving electromagnetic resilience						
Also to ensure that good electromagnetic engineering practices are employed (see 2.3.25) as appropriate during installation and commissioning.	HR	HR	M	M	2.4.4	
Also to help ensure that the safety-related system will comply with the EMC test specifications from 2.1.3 and (if appropriate) 2.1.4 during verification and validation (see 2.5.2).						
Other technique or measure used for implementation, integration, installation and commissioning:	-	-	-	-		
Other technique or measure used for implementation, integration, installation and commissioning:	-	-	-	-		
Other technique or measure used for implementation, integration, installation and commissioning:	-	-	-	-		
Techniques and measures for verification and validation (including testing)					2.5	

Overview of some electromagnetic resilience techniques and measures	Importance				References in this Code of Practice	Applied: Yes/No Add any comments, references, and links
	SIL1	SIL2	SIL3	SIL4		
Applying verification, and validation techniques and measures						
To verify and/or validate as far as is practicable that the design techniques and measures that have been applied function according to the relevant design specification (created by 2.1).	M	M	M	M	2.5.1	
(Note that EMC testing is covered in 2.5.2 and 2.5.3.)						
Verification testing to the EMC test plan resulting from 2.1.3 (and 2.1.4 if appropriate)	M	M	M	M	2.5.2	
Using non-standardised ad hoc checks or tests						
To help ensure that the safety-related system or any component part of it has sufficient electromagnetic resilience, taking into account the final safety integrity level/systematic capability be ng aimed for.	HR	HR	HR	HR	2.5.3	
Verifying correct installation and commissioning						
Having regard to the constraints and additional measures listed as the result of applying 2.4.1 and 2.4.2, and any others not listed in those sections.	HR	HR	HR	M	2.5.4	
Other technique or measure used for verification and validation (including testing):	-	-	-	-		
Other technique or measure used for verification and validation (including testing):	-	-	-	-		
Other technique or measure used for verification and validation (including testing):	-	-	-	-		
Other technique or measure used for verification and validation (including testing):	-	-	-	-		
Techniques and measures for maintenance, refurbishment, repair, modification, upgrade, etc., over the lifecycle					2.6	
Assessment of changes in the electromagnetic environment						
And if necessary modify/upgrade as required so that availability is maintained at a high level (discussed in 2.1.3).	HR	HR	HR	M	2.6.1	

Overview of some electromagnetic resilience techniques and measures	Importance				References in this Code of Practice	Applied: Yes/No Add any comments, references, and links
	SIL1	SIL2	SIL3	SIL4		
Assessment of continuing correct installation						
Having regard to the constraints and additional measures listed as the result of applying 2.4.1, and any others not listed in that section.	HR	HR	HR	M	2.6.2	
Maintaining electromagnetic resilience despite modifications or changes						
Assessing proposed changes to the safety-related system to ensure that repairs, modifications, upgrades, refurbishment, etc. do not unacceptably degrade its electromagnetic resilience.	HR	HR	M	M	2.6.3	
Other technique or measure used for maintenance, refurbishment, repair, modification, upgrade, etc., throughout the lifecycle:	-	-	-	-		
Other technique or measure used for maintenance, refurbishment, repair, modification, upgrade, etc., throughout the lifecycle:	-	-	-	-		
Other technique or measure used for maintenance, refurbishment, repair, modification, upgrade, etc., throughout the lifecycle:	-	-	-	-		
Other technique or measure used for maintenance, refurbishment, repair, modification, upgrade, etc., throughout the lifecycle:	-	-	-	-		
Other technique or measure used for maintenance, refurbishment, repair, modification, upgrade, etc., throughout the lifecycle:	-	-	-	-		
Techniques and measures for decommissioning					2.7	
To ensure – where appropriate – that decommissioning does not cause unacceptable functional safety risks due to the electromagnetic resilience of its safety-related system being degraded by the decommissioning process.	HR	HR	HR	M	2.7	
Other technique or measure used for decommissioning:	-	-	-	-		

Overview of some electromagnetic resilience techniques and measures	Importance				References in this Code of Practice	Applied: Yes/No Add any comments, references, and links
	SIL1	SIL2	SIL3	SIL4		
Other technique or measure used for decommissioning:	-	-	-	-		
Other technique or measure used for decommissioning:	-	-	-	-		
Techniques and measures for integrating third-party items into safety-related systems					2.8	
The general iterative approach As shown in Figure 2.1.					2.8.1	
Suppliers' certifications and electromagnetic performance Suppliers' markings, certifications and declarations (including CE marking with respect to the EMC Directive) should not be taken as reliable evidence of electromagnetic performance.					2.8.2	
Alternatively, use custom-manufactured elements And make producing reliable evidence of electromagnetic performance part of the contract specification.					2.8.3	
Other technique or measure used for integrating third-party items into a safety-related system:						
Other technique or measure used for integrating third-party items into a safety-related system:						
Other technique or measure used for integrating third-party items into a safety-related system:						

ANNEX B

References

Annex B.1   General references

[1]   IEC 61508-7 Ed.2:2010 *Functional safety of electrical/electronic/ programmable electronic safety related systems - Part 7: Overview of techniques and measures*, an IEC Basic Safety Standard (2010), http://webstore.iec.ch

[2]   IEC 61000-1-2:2016 *Electromagnetic compatibility (EMC) - Part 1-2: General - Methodology for the achievement of functional safety of electrical and electronic systems including equipment with regard to electromagnetic phenomena*, an IEC Basic Safety Standard (2016), http://webstore.iec.ch

[3]   IEC 61326 *Part 3-1: Immunity requirements for safety-related systems and for equipment intended to perform safety-related functions (functional safety) – General industrial applications*, http://webstore.iec.ch

[4]   IEC 61326 *Part 3-2: Immunity requirements for safety-related systems and for equipment intended to perform safety-related functions (functional safety) – Industrial applications with specified electromagnetic environment*, http://webstore.iec.ch

[5]   IEC 61000-6-7:2014 *Electromagnetic compatibility (EMC) – Part 6-7: Generic standards – Immunity requirements for systems, equipment and products intended to perform functions in a safety-related system (functional safety) in industrial environments*, http://webstore.iec.ch

[6]   Guides on EMC for Functional Safety published by the IET in 2008 and 2013, from http://www.theiet.org/factfiles/emc/

[7]   2014/30/EU, the European Union's Directive on EMC, http://eur-lex. europa.eu/legal-content/EN/TXT/?uri=CELEX:32014L0030.

[8]   IEC 61508-1 Ed.2:2010 *Functional safety of electrical/electronic/ programmable electronic safety related systems – Part 1: General Requirements*, IEC Basic Safety Publication (2010), http://webstore.iec.ch

[9]   Defense Acquisition Regulations System, Department of Defense. 48 CFR Parts 202, 231, 244, 246, and 252, RIN 0750–AH88, Defense Federal Acquisition Regulation Supplement: *Detection and Avoidance of Counterfeit Electronic Parts* (DFARS Case 2012–D055), published in the Federal Register / Vol. 79, No. 87 / Tuesday, May 6, 2014 / Rules and Regulations.

[10]   UKAS document LAB 34 *The expression of measurement uncertainty in EMC testing*, 2002: www.ukas.com/download/publications/publications-relating-to-laboratory-accreditation/Lab34.pdf

[11]   *Why EMC Immunity Testing is Inadequate for Functional Safety*, Keith Armstrong, IEEE 2004 International Symposium on EMC, Santa Clara, CA, August 9-13, ISBN: 0-7803-8444-X

[12]   *Why increasing immunity test levels is not sufficient for high-reliability and critical equipment*, Keith Armstrong, IEEE 2009 International Symposium on EMC, Austin, TX, August 17-21, ISBN: 978-1-4244-4285-0

[13]   *Testing for immunity to simultaneous disturbances and similar issues for risk managing EMC*, Keith Armstrong, IEEE 2012 International Symposium on EMC, Pittsburgh, PA, August 5-10, ISBN: 978-1-4673-2059-7

[14]   *Why Do We Need an IEEE EMC Standard on Managing Risks?*, Keith Armstrong, 2016 IEEE Electromagnetic Compatibility Magazine – Volume 5 – Quarter 1, pages 81-84, http://ieeexplore.ieee.org/stamp/stamp.jsp?arnumber=7477140

Annex B.2   Good EMC engineering for systems and installations

[20]   IEC 61000-5-2 ed 1.0, November 1997 *Electromagnetic compatibility (EMC) – Part 5: Installation and mitigation guidelines – Section 2: Earthing and cabling*, http://webstore.iec.ch

[21]   *Good EMC Engineering Practices in the Design and Construction of Industrial Cabinets (relevant for all types of electrical/electronic equipment)*, Keith Armstrong, REO (UK) Ltd., www.reo.co.uk/technical_resources

[22]   *Good EMC Engineering Practices in the Design and Construction of Fixed Installation*, Keith Armstrong, REO (UK) Ltd., www.reo.co.uk/technical_resources

[23]   *EMC for Systems and Installations*, Tim Williams and Keith Armstrong, Newnes 2000, ISBN 0-7506-4167-3, www.bh.com/newnes, RS Components Part No. 377-6463

[24]   IEC/TR 61000-5-6 *Electromagnetic Compatibility (EMC) – Part 5: Installation and mitigation guidelines – Section 6: Mitigation of external influences*, http://webstore.iec.ch

[25]   *Designing Electronic Systems for EMC*, William G Duff, 2001, ISBN: 978-1-891121-42-5, Scitech Publishing, Inc., www.scitechpublishing.com

[26]   *Complying with IEC 61800-3 – Good EMC Engineering Practices in the Installation of Power Drive Systems*, Keith Armstrong, REO (UK) Ltd., www.reo.co.uk/technical_resources

[27]   *Mains Harmonics (problems and solutions)*, Keith Armstrong, REO (UK) Ltd., www.reo.co.uk/technical_resources

[28]   *Power Quality (problems and solutions)*, Keith Armstrong, REO (UK) Ltd., www.reo.co.uk/technical_resources

[29]   *Grounds for Grounding*, Elya B Joffe and Kai-Sang Lock, IEEE Press, John Wiley & Sons, Inc., 2010, ISBN 978-04571-66008-8

[30]   *Protection of Electronics in High-Power Installations: Theory, Guidelines and Demonstrations*, P C T van der Laan and A P J van Duerson (Eindhoven University of Technology), CIGRÉ Symposium, Lausanne, 1993, paper 600-08

[31]   *Reliable Protection of Electronics Against Lightning: Some Practical Examples*, P C T van der Laan and A P J van Duerson (Eindhoven University), IEEE Trans. EMC, Vol 40, No 4, November 1998, pp 513-520

[32]   *Design Philosophy for Grounding*, M A van Houten and P C T van der Laan (Eindhoven University of Technology), Proc. 5th Int. Conf. on EMC, York, UK, IERE Publication No. 71 (1986) p 267-272

[33]   *Protection of Cables by Open-Metal Conduits*, S Kapora, E Laermans, A P J van Duerson, IEEE Trans. EMC, Vol. 52, No. 4, Nov. 2010, pp 1026 – 1033

[34]   *Analysis of Electromagnetic Shielding of Cables and Connectors (keeping currents/voltages where they belong)*, Lothar O. (Bud) Hoeft, PhD, IEEE, 2002, http://simbilder.com/ieee/34/EMag_Shielding_of_Cables_and_Connectors.pdf

[35]   IEC 364-4-444:1996 *Electrical Installations of Buildings – Part 4: Protection for safety – Chapter 44: Protection against overvoltages- Section 444: Protection against electromagnetic interference (EMI) in installations of buildings*, http://webstore.iec.ch

[36]   IEC 62305 *Protection against lightning*;

Part 1: *General Principles*

Part 2: *Risk Management*

Part 3: *Physical damage to structures and life hazard*

Part 4: *Electrical and electronic systems within structures connected to telecommunications and signalling networks – Performance requirements and testing methods*

[37]   *Combined Effects of Several, Simultaneous, EMI Couplings*, Michel Mardiguian, 2000 IEEE International Symposium on EMC, Washington D.C., August 21-25 2000, ISBN 0-7803-5680-2, pp. 181-184

[38]   IEC/TR 61000-1-5, *High power electromagnetic (HPEM) effects on civil systems*

[39]   *The Development of High-Power Electromagnetic (HPEM) Publications in the IEC: History and Current Status*, Dr. William A. Radasky, IEEE EMC Society Newsletter, Issue No. 216, Winter 2008, www.ieee.org/organizations/pubs/newsletters/emcs/winter08/hpem.html

[40]   IEC/TR 61000-5-3 *Installation and mitigation guidelines – HEMP protection concepts*

[41]   IEC/TS 61000-5-4 *Installation and mitigation guidelines – Immunity to HEMP – Specifications for protective devices against HEMP radiated disturbance*

[42]   IEC 61000-5-5 *Installation and mitigation guidelines – Specification of protective devices for HEMP conducted disturbance*

[43]   IEC 61000-5-8 *HEMP protection measures for the distributed infrastructure*

[44]   IEC 61000-5-9 *Installation and mitigation guidelines – System-level susceptibility assessments for HEMP and HPEM.*

[45]   ORNL/Sub/91-SG9131/1:1992 *Recommended engineering practice to enhance the EMI/EMP immunity of electric power systems*, Oak Ridge National Laboratory, USA.

Annex B.3   Good EMC engineering for individual items of equipment

[60]   *EMC for Printed Circuit Boards – Basic and Advanced Design and Layout Techniques*, Second Edition, Nutwood UK December 2010, ISBN 978-0-9555118-5-1, (the 2nd Edition is identical to the 1st Edition except for the book's format), www.emcacademy.org/books.asp

[61]   *EMC Design Techniques for Electronic Engineers*, Keith Armstrong, Armstrong/Nutwood UK 2010, ISBN: 978-0-9555118-4-4, www.emcacademy.org/books.asp

[62]   *EMC for Product Designers*, 4th Edition, Tim Williams, Newnes, December 2006, ISBN: 0-750-68170-5

[63]   *High Speed Digital Design: A Handbook Of Black Magic*, Johnson, Howard and Graham, Martin, Prentice Hall, 1993, ISBN 0-13-39-5724-1

[64] *Robust Electronic Design Reference Book*, Volumes I and II, John R Barnes, Kluwer Academic Publishers, 2004, ISBN: 1-4020-7739-4

[65] *Printed Circuit Board Design Techniques for EMC Compliance*, Second Edition, A Handbook for Designers, M Montrose, IEEE Press 2000, ISBN 0-7803-5376-5, http://www.ieee.org/ieestore

[66] *EMC and the Printed Circuit Board - Design, Theory and Layout Made Simple*, M Montrose, IEEE Press 1998, ISBN 0-7803-4703-X, http://www.ieee.org/ieestore

[67] *Electromagnetic Compatibility Engineering*, Henry W. Ott, John Wiley & Sons, 2009, ISBN: 978-0-470-18930-6

[68] *Ageing of Shielding Joints, Shielding Performance and Corrosion*, Lena Sjögren and Mats Bäckström, IEEE EMC Society Newsletter, Summer 2005, www.ieee.org/organizations/pubs/newsletters/emcs/summer05/practical.pdf

[69] *Improving the shielding effectiveness of a board-level shield by bonding it with the waveguide-below-cutoff principle* A Degraeve, D. Pissoort, K. Armstrong, 10th International IEEE Workshop on the Electromagnetic Compatibility of Integrated Circuits (EMC Compo), Edinburgh, UK, 2015

Annex B.4   Software design techniques and measures

[100] *Software Engineering for Real Time Systems*, J E Cooling, Pearson Education2003, ISBN 0201596202

[101] *Dependability of Computer Systems*, EWICS Technical Committee 7, Elsevier Applied Science1989 ISBN 1851663819

[102] "Defensive Programming", http://www.princeton.edu/~achaney/tmve/wiki100k/docs/Defensive_programming.html

[103] *Safety Critical Systems Handbook: A Straightforward Guide to Functional Safety*, IEC 61508 (2010 Edition) and Related Standards, Including: Process IEC 61511, Machinery IEC 62061

[104] NASA Software Safety Guidebook from: www.fmeainfocentre.com/handbooks/nasasoftwareguidbook.doc

[105] National Conference on Nonlinear Systems & Dynamics, NCNSD-2003, *N-Version programming method of Software Fault Tolerance: A Critical Review*, Bharathi V, http://ncnsd.org/proceedings/proceeding03/html/pdf/173-176.pdf

[106] *Formal Methods in Safety-Critical Standards*, Jonathan Bowen, Oxford University Computing Laboratory, 11 Keble Road, Oxford OX1 3QD, UK. http://reference.kfupm.edu.sa/content/f/o/formal_methods_in_safety_critical_standa_100249.pdf

[107] *Using Software Protocols to Mask CAN BUS Insecurities*, B R Kirk, IEE Colloquium on the Electromagnetic Compatibility of Software, Thursday, Savoy Place, London, WC2R OBL, 12 November 1998, IEE document reference 98/471, available from the IEE Library at Savoy Place, libdesk@theiet.org, or archives@theiet.org, telephone 020 7344 8407, fax: 020 7344 846.

[108] *Profibus specification – Profisafe – Profile for Safety Technology*, Version 1.30, June 2004, Profibus International

[109]   IEC 61784-3 Ed.1 CDV *Digital data communications for measurement and control: Part 3: Profiles for functional safety communications in industrial networks*

[110]   *32-Bit Cyclic Redundancy Codes for Internet Applications,* Philip Koopman, *International Conference on Dependable Systems and Networks,* 2002

[111]   Cyclic redundancy check (CRC), http://en.wikipedia.org/wiki/Cyclic_redundancy_check

CRC is an error-detecting code commonly used in digital networks and storage devices to detect accidental changes to raw data. Blocks of data entering these systems get a short check value attached, based on the remainder of a polynomial division of their contents; on retrieval the calculation is repeated, and corrective action can be taken against presumed data corruption if the check values do not match.

CRCs are so called because the check (data verification) value is a redundancy (it expands the message without adding information) and the algorithm is based on cyclic codes. CRCs are popular because they are simple to implement in binary hardware, easy to analyse mathematically, and particularly good at detecting common errors caused by noise in transmission channels. Because the check value has a fixed length, the function that generates it is occasionally used as a hash function.

The CRC was invented by W. Wesley Peterson in 1961; the 32-bit polynomial used in the CRC function of Ethernet and many other standards is the work of several researchers and was published during 1975.

[112]   Error Correction: www.wikipedia.org/wiki/Error_correction

[113]   The avionics standard based on the concept of partitioning the processor time, memory ranges and I/O access: http://en.wikipedia.org/wiki/ARINC_653, also: "ARINC 653 An Avionics Standard for Safe, Partitioned Systems", www.computersociety.it/wp-content/uploads/2008/08/ieee-cc-arinc653_final.pdf.

[114]   An operating system that supports partitioning: www.ghs.com/products/safety_critical/arinc653.html

[115]   Another operating system that supports partitioning: www.windriver.com/products/platforms/safety_critical_arinc_653/

[116]   A paper describing the concepts of partitioning operating systems: http://air.di.fc.ul.pt/air-ii/downloads/27th-DASC-Paper.pdf

[117]   Reliable/redundant array of independent/inexpensive nodes (RAIN): http://en.wikipedia.org/wiki/Reliable_array_of_independent_nodes

RAIN is an architectural approach to computing and network-attached computer storage (or NAS), that combines commodity or low-cost computing hardware with management software to address the reliability and availability shortcomings of non-redundant NAS systems.

[118]   IEC 61508-3 Ed.2:2010 *Functional safety of electrical/electronic/programmable electronic safety-related systems – Part 3: Software requirements,* IEC Basic Safety Publication (2010), http://webstore.iec.ch

[119]   IEC 61508-2 Ed.2:2010 *Functional safety of electrical/electronic/programmable electronic safety related systems – Part 2: Requirements for electrical/electronic/programmable electronic safety related systems,* IEC Basic Safety Publication (2010), http://webstore.iec.ch

[120]   IEEE Standard for Floating-Point Arithmetic, 754-2008, from http://ieeexplore.ieee.org/xpl/mostRecentIssue.jsp?punumber=4610933

[121] *Synthesizing optimal fixed-point arithmetic for embedded signal processing*, Hass, K.J., 53rd IEEE International Midwest Symposium on Circuits and Systems (MWSCAS), 1-4 Aug. 2010, Seattle, WA, pp 61 – 64, ISBN: 978-1-4244-7771-5

[122] *System Software Support of Hardware Efficiency*, by Thomas Kaegi and Igor Schagaev, an eBook from: www.it-acs.co.uk/book.html

[123] RTCA DO-178C North American Avionics Software, *Software Considerations in Airborne Systems and Equipment Certification*

[124] *Safe Program Execution with Diversified Encoding*, Martin Susskraut *et al*, Embedded World 2015, www.embedded-world.eu

[125] *A Time & Space Partitioned DO-178 Level A Certifiable RTOS, Supports x86, PowerPC, ARM and MIPS processors*, www.ddci.com/products_deos.php

[126] Visit: www.qnx.com/solutions/industries/defense.html

[127] *Comparison of the software requirements in safety related cases*, according to IEC 61508, by Sigita Andrulyte, Josef Börcsök, www.wseas.us/e-library/conferences/2013/Budapest/CSECS/CSECS-33.pdf

[128] *Liveness*, https://en.wikipedia.org/wiki/Liveness

[129] *The Ariane 5 accident,* http://sunnyday.mit.edu/accidents/Ariane5 accidentreport.html

[130] *The Mars Climate Orbiter failure*, https://en.wikipedia.org/wiki/Mars_Climate_Orbiter

[131] *Error Correction Coding: Mathematical Methods and Algorithms*, Todd K Moon, Wiley 2005, ISBN: 0-471-648-00-0.

[132] "Secure Hashing", CSRC.NIST.gov/groups/ST/toolkit/secure_hashing.html

[133] *Software-Based Memory Testing*, Barr, Michael, Embedded Systems Programming, July 2000, pp. 28-40.

[Note that there are some comments in response to this publication which you can find at the end of it on the website: http://www.barrgroup.com/Embedded-Systems/How-To/Memory-Test-Suite-C]

[134] *Data caching*, see https://en.wikipedia.org/wiki/Cache

[135] *DDR2 Synchronous Dynamic Data Interface*, https://en.wikipedia.org/wiki/DDR2_SDRAM

[136] *DDR3 Synchronous Dynamic Data Interface*, https://en.wikipedia.org/wiki/DDR3_SDRAM

[137] *JTAG101 – IEEE 1149.x and Software Debug*, Randy Johnson, Steward Christie (Intel Corp.2009),

[138] IEEE 1149.6, *A Boundary-Scan Standard for Advanced Digital Networks*

[139] *Nonlinear Signal Processing: a Statistical Approach*, G. R. Arce, Wiley New Jersey November 2004, Print ISBN: 978-0-471-67624-9, Online ISBN: 978-0-471-69185-3

[140] CEPT ECC Rec 70-03, *ERC Recommendation 70-03 Relating to the use of Short Range Devices (SRD)*, www.erodocdb.dk/docs/doc98/official/pdf/rec7003e.pdf

[141] Exception handling, https://en.wikipedia.org/wiki/Exception_handling

[142] IEEE Transactions on Dependable and Secure Computing (TDSC), www.computer.org/web/tdsc/about

[143] *The Embedded Reliable Processing System (TERPS) — A Robust Architecture that Achieves Forward Progress in Near-Continuous Electromagnetic Interference*, Cagdas Dirik, Amol Gole, Samuel Rodriguez, Hongxia Wang, and Bruce Jacob, Electrical & Computer Engineering Dept., University of Maryland — College Park, www.ece.umd.edu/~blj • blj@umd.edu, Technical Report UMD-SCA-2004-10-01 — November 2004

[144] *Microprocessor Based Protection Systems*, Churchley, Andrew (1991-11-30), Springer. p.64. ISBN 9781851666119, https://books.google.co.uk/books?id=vNzWLxmzuUsC&pg=PA64&redir_esc=y&hl=en#v=onepage&q&f=false

[145] A microprocessor intentionally designed without any interrupts: https://en.wikipedia.org/wiki/VIPER_microprocessor.

Annex B.5 IEC and CISPR standardised EMC test methods

Note: *This is not an exhaustive list, and new standards are constantly being created.*

[200] IEC 61000-4-2 *Immunity to personnel electrostatic discharge (ESD)*

[201] IEC 61000-4-3 *Immunity to continuous radio-frequency radiation using an anechoic chamber*

[202] IEC 61000-4-4 *Immunity to electrical fast transients and bursts (EFT/B)*

[203] IEC 61000-4-5 *Immunity to surges*

[204] IEC 61000-4-6 *Immunity to continuous conducted radio-frequency currents*

[205] IEC 61000-4-8 *Immunity to power-frequency magnetic fields*

[206] IEC 61000-4-9 *Immunity to pulsed magnetic fields*

[207] IEC 61000-4-10 *Immunity to damped oscillatory magnetic fields*

[208] IEC 61000-4-11 *Immunity to voltage dips, dropouts, short interruptions and voltage variations*

[209] IEC 61000-4-12 *Immunity to ring wave surges*

[210] IEC 61000-4-13 *Immunity to distorted AC supply waveforms up to 2kHz*

[211] IEC 61000-4-14 *Immunity to AC supply voltage fluctuations*

[212] IEC 61000-4-16 *Immunity to conducted common-mode disturbances DC-150kHz*

[213] IEC 61000-4-17 *Immunity to voltage ripple on DC electrical power supplies*

[214] IEC 61000-4-18 *Immunity to damped oscillatory surges*

[215] IEC 61000-4-19 (draft) *Immunity to conducted differential mode disturbances 2-150kHz*

[216] IEC 61000-4-20 *Immunity to continuous radio-frequency radiation using a TEM Cell*

[217] IEC 61000-4-21 *Immunity to continuous RF radiation using a Reverberation Chamber*

[218] IEC 61000-4-25 *Immunity to HEMP for equipment and systems*

[219] IEC 61000-4-27 *Immunity to unbalance in three-phase AC power supplies*

[220] IEC 61000-4-28 *Immunity to variations in AC power supply frequency*

[221] IEC 61000-4-31 *Immunity to conducted broadband noise*

[222] IEC 61000-4-34 *Immunity to supply voltage dips, dropouts and voltage variations for equipment consuming more than 16A per phase*

[223]   IEC 61000-4-36 *Immunity to Intentional EMI*

[224]   IEC 61000-4-39 (future) *Measuring methods for radiation sources in close proximity, 9kHz to 6GHz. (Note that this standard tests Far-Fields, not Near-Fields, see [317]).*

[225]   C37.90.1-2002 *IEEE Standard for Surge Withstand Capability (SWC) Tests for Relays and Relay Systems Associated with Electric Power Apparatus*

[226]   IEC 61000-6-1 *EMC Generic standards — Immunity for residential, commercial and light-industrial environments*

[227]   IEC 61000-6-2 *EMC Generic standards — Immunity for industrial environments*

[228]   IEC 61000-6-3 *EMC Generic standards - Emission standard for residential, commercial and light-industrial environments*

[229]   IEC 61000-6-4 *EMC Generic standards — Emission standard for industrial environments*

[230]   IEC/TS 61000-6-5 *EMC Generic standards — Immunity for power station and substation environments*

[231]   IEC 61000-4-33 *Testing and measurement techniques — Measurement methods for high-power transient parameters*

[232]   IEC 61000-4-35 *Testing and measurement techniques — HPEM simulator compendium*

[232]   IEC 61000-4-23 *Test methods for protective devices for HEMP and other radiated disturbances*

[233]   IEC 61000-4-24 *Test methods for protective devices for HEMP conducted disturbances*

[234]   IEC 61000-6-6 *HEMP immunity for indoor equipment*

[235]   IEC 62561 *Lightning protection system components (LPSC)*

[236]   IEC 61000-4-32 *HEMP simulator compendium*

[237]   CISPR 11 *Industrial, scientific and medical equipment — Radio-frequency disturbance characteristics — Limits and methods of measurement*

[238]   CISPR 13 *Sound and television broadcast receivers and associated equipment — Radio-frequency disturbance characteristics — Limits and methods of measurement*

[239]   CISPR 14-1 *Electromagnetic compatibility — Requirements for household appliances, electric tools and similar apparatus — Part 1: Emission*

[240]   CISPR 14-2 *Electromagnetic compatibility — Requirements for household appliances, electric tools and similar apparatus — Part 2: Immunity — Product family standard*

[241]   CISPR 15 *Limits and methods of measurement of radio disturbance characteristics of electrical lighting and similar equipment*

[242]   CISPR 22 *Information technology equipment of radio disturbance characteristics — Limits and methods of measurement*

[243]   CISPR 32 *Electromagnetic compatibility of multimedia equipment — Emission requirements*

[244]   CISPR 24 *Information technology equipment — Immunity characteristics — Limits and methods of measurement*

Annex B.6   Automotive industry EMC test standards

Note: *This is not an exhaustive list, and new standards are constantly being created.*

[300]   CISPR 12 *Vehicle emissions measurements for the protection of off-board radio communications*

[301]   CISPR 25 *Vehicles, boats and internal combustion engines — Radio disturbance characteristics - Limits and methods of measurement for the protection of on-board receivers*

[302]   ISO 10605 *Immunity to electrostatic discharge (now mostly replaced by IEC 61000-4-2)*

[303]   ISO 7637-2 *Immunity of power lines to conducted transients*

[304]   ISO 7637-3 *Immunity of signal, data and control lines to conducted transients*

[305]   ISO 11451-1 *Road vehicles -- Vehicle test methods for electrical disturbances from narrowband radiated electromagnetic energy -- Part 1: General principles and terminology*

[306]   ISO 11451-2 *Off-vehicle radiation sources*

[307]   ISO 11451-3 *On-board radio communications equipment*

[308]   ISO 11451-4 *Bulk current injection (BCI)*

[309]   ISO 11452-1 *Road vehicles -- Component test methods for electrical disturbances from narrowband radiated electromagnetic energy -- Part 1: General principles and terminology*

[310]   ISO 11452-2 *Absorber-Lined Shielded Enclosure (ALSE)*

[311]   ISO 11452-3: *Transverse Electromagnetic (TEM) cell*

[312]   ISO 11452-4 *Bulk Current Injection (BCI)*

[313]   ISO 11452-5 *Stripline*

[314]   ISO 11452-6 *Parallel plate antenna*

[315]   ISO 11452-7 *Direct RF Power Injection (DPI)*

[316]   ISO 11452-8 *Immunity to magnetic fields*

[317]   ISO 11452-9.2 *Road vehicles — Component test methods for electrical disturbances from narrowband radiated electromagnetic energy — Part 9: Portable transmitters*

[This is a Near-Field radiated immunity test based upon the Ford Motor Company's test method RI115 "RF Immunity to hand portable transmitters" in their EMC-CS-2009.1, "EMC Specification For Electrical/Electronic Components and Subsystems".

Many EMC test labs around the world are equipped for, and familiar with doing this test.]

[318]   Ford Motor Company: EMC-CS-2009.1, *EMC Specification For Electrical/Electronic Components and Subsystems.*

Annex B.7   Marine industry EMC test standards

Note: *This is not an exhaustive list, and new standards are constantly being created.*

[350]   IEC 60945 *Maritime navigation and radiocommunication equipment and systems – General requirements – Methods of testing and required test results*

[351]   IEC 60533 *Electrical and electronic installations in ships – Electromagnetic compatibility (EMC) – Ships with a metallic hull*

[352]   IEC 60092–201 *Electrical installations in ships — Part 201: System design — General*

Annex B.8   Undersea industry EMC test standards

Note: *This is not an exhaustive list, and new standards are constantly being created.*

[360]   ISO 13628-1 *Petroleum and natural gas industries -- Design and operation of subsea production systems -- Part 1: General requirements and recommendations*

[361]   ISO 13628-6 *Petroleum and natural gas industries -- Design and operation of subsea production systems -- Part 6: Subsea production control systems*

[362]   API 17A *Recommended Practice 17A, Design and Operation of Subsea Production Systems*

[363]   API 17F *Specification for Subsea Production Control Systems - Petroleum and natural gas industries - Design and operation of subsea production systems -Part 6: Subsea production control systems*

Annex B.9   Rail industry EMC standards and guidance documents

Note: *This is not an exhaustive list, and new standards are constantly being created.*

[400]   EN50121-series Railway Applications – Electromagnetic Compatibility

• EN 50121-1 Railway applications – Electromagnetic Compatibility Part 1 General

• EN 50121-2 Railway applications – Electromagnetic Compatibility Part 2 Emissions of the whole railway system to the outside world

• EN 50121-3-1 Railway applications – Electromagnetic Compatibility Part 3-1 Rolling Stock Train & complete vehicle

• EN 50121-3-2 Railway applications – Electromagnetic Compatibility Part 3-2 Rolling Stock Apparatus

• EN 50121-4 Railway applications – Electromagnetic Compatibility Part 4 Emission and Immunity of signalling and telecommunications apparatus

• EN 50121-5 Railway applications – Electromagnetic Compatibility Part 5 Emission and Immunity of fixed power supply installations and apparatus

[401]   EN 50155 *Railway applications — Electronic equipment used on rolling stock*

[402]   NR/L1/SIG/30040 *EMC Strategy for Network Rail*

[403]   NR/L2/RSE/30041 *EMC Assurance Process for Network Rail*

[404]   Railway Group Standard GE/RT8270 *Assessment of Compatibility of Rolling Stock and Infrastructure*

[405]   Railway Group Standard GE/RT8015 *Electromagnetic Compatibility Between Railway Infrastructure and Trains*, to be replaced by GE/RT8076 *Electromagnetic Compatibility of Train Detection Infrastructure with Rail Vehicles*

[406]   LUL Category 1 Standard S1222 *Electromagnetic Compatibility (EMC)*

[407]   LUL Guidance Document G222 *EMC best practice*

[408]   LUL Category 1 Standard S1193 *Electromagnetic Compatibility with LU Signalling System Assets*

[409]   EN 50592 *Railway Applications – Testing of rolling stock for electromagnetic compatibility with axle counters*

[410]   BS EN 50238-1 *Incorporating corrigenda May 2010 and November 2014 Railway applications – Compatibility between rolling stock and train detection systems*

[411]   PD CLC/TS 50238-2 *Railway applications – Compatibility between rolling stock and train detection systems Part 2: Compatibility with track circuits*

[412]   CLC/TS 50238-3 *Railway applications – compatibility between rolling stock and train detection systems – Part 3 Compatibility with axle counters*

[413]   UNISIG ERTMS/ETCS FFFIS for Eurobalise REF:SUBSET-036 [contains in-band transient susceptibility test]

[414]   NR/SP/SIG/50002, *Methodology for the demonstration of compatibility with single rail Reed Track Circuits on the AC railway*

[415]   NR/GN/SIG/50003 *Methodology for the demonstration of compatibility with Double Rail Reed Track Circuits on the DC railway*

[416]   NR/SP/SIG/50004 *Methodology for the demonstration of electrical compatibility with DC (AC-immune) Track Circuits*

[417]   NR/SP/SIG/50005 *Methodology for the demonstration of electrical compatibility with 50 Hz single rail track circuits*

[418]   NR/SP/SIG/50006 *Methodology for the demonstration of electrical compatibility with 50 Hz double rail track circuits*

[419]   NR/SP/SIG/50007 *Methodology for the demonstration of compatibility with HVI Track Circuits*

[420]   NR/SP/SIG/50008 *Methodology for the demonstration of compatibility with TI21 Track Circuits*

[421]   NR/SP/SIG/50009 *Methodology for the demonstration of compatibility with FS2600 Track Circuits*

[422]   NR/GN/SIG/50010 *Methodology for the demonstration of compatibility with train detection systems on non-electrified railways*

[423]   NR/SP/SIG/50011 *Methodology for the Demonstration of Compatibility with Axle Counters*

[424]   NR/SP/SIG/50012 *Methodology for the Demonstration of Compatibility with TPWS Track Sub-system*

[425]   NR/SP/SIG/50013 *Methodology for the Demonstration of Compatibility with Interlockings*

[426]   NR/GN/SIG/50014 *Methodology for the Demonstration of Compatibility with Lineside Equipment*

[427]   NR/GN/SIG/50018 *Methodology for the Demonstration of Compatibility with Neighbouring Railways*

[428]   NR/L2/TEL/31107 *Limits and Test Method of induced voltages on telecommunications cables due to electrification systems*

[429]   EN 50617-1 *Railway applications – Technical parameters of train detection systems for the interoperability of the trans-European railway system Part 1: Track circuits*

[430]   EN 50617-2 *Railway Applications – Technical parameters of train detection systems for the interoperability of the trans-European railway system Part 2: Axle counters - CORR: February 29, 2016*

Annex B.10   Civilian avionics and aerospace industry EMC test standards

Note: *This is not an exhaustive list, and new standards are constantly being created.*

[460]   RTCA DO-160 *Environmental Conditions and Test Procedures for Airborne Equipment, Section 22: Lightning induced transient susceptibility, Section 23: Lightning direct effects*

[461]   RTCA DO-160 *Environmental Conditions and Test Procedures for Airborne Equipment, Section 20: Radio frequency susceptibility (Radiated and Conducted)*

[462]   RTCA DO-160 *Environmental Conditions and Test Procedures for Airborne Equipment, Section 16: Power Input*

[463]   RTCA DO-294C *Guidance on Allowing Transmitting Portable Electronic Devices (T-PEDS) on Aircraft*

[464]   RTCA DO-307 *Aircraft Design and Certification for Portable Electronic Device (PED) Tolerance*

[465]   ED-130 *Guidance for the Use of Portable Electronic Devices (PEDS) On Board Aircraft*

Annex B.11   Military industry EMC test standards

Note: *This is not an exhaustive list, and new standards are constantly being created.*

[500]   Def-Stan 59-411 *Part 1: EMC Management and Planning*

[501]   Def-Stan 59-411 *Part 2: The Electric, Magnetic and Electromagnetic Environment*

[502]   Def-Stan 59-411 *Part 3: EMC Test Methods and Limits for Equipment and Sub Systems*

[503]   Def-Stan 59-411 *Part 4: EMC Platform and System Test and Trials*

[504]   Def-Stan 59-411 *Part 5: EMC Code of Practice for Tri-Service Design and Installation*

[505]   MIL-STD-461 *Requirements for the Control of Electromagnetic Interference Characteristics of Subsystems and Equipment*

[506]   MIL-STD-464 *Electromagnetic Environmental Effects Requirements for Systems*

[507]   AECTP-250 *Electrical and Electromagnetic Environmental Conditions*

[508]   AECTP-500 *Electromagnetic Environmental Effects Test and Verification*

Annex B.12   ITE, Telecommunications and Wireless industry EMC test standards

Note: *This is not an exhaustive list, and new standards are constantly being created.*

[550]   EN 50310 *Application of equipotential bonding and earthing at premises with information technology equipment*, http://shop.bsigroup.com/en

[551]   ETSI EN 300 253 *Earthing and bonding of telecommunication equipment in telecommunication centres*, www.etsi.org/deliver/etsi_en/300200_300299/3002 53/02.01.01_60/en_300253v020101p.pdf

[552]   ITU-T Recommendation K.27 *Bonding configurations and earthing within a elecommunications building*, www.itu.int/rec/T-REC-K.27-199605-I

[553] ITU Recommendation K.35 *Bonding configurations and earthing at remote electronic sites*, www.itu.int/rec/T-REC-K.35-199605-I

[554] EN 50174-2 *Information Technology – Cabling Installation Part 2: Installation planning and practice inside buildings*, http://shop.bsigroup.com/en

[555] ITU-T K.78 *High altitude electromagnetic pulse immunity guide for telecommunication centres*

[556] ITU-T Handbook, *The Protection of Telecommunication Lines and Equipment Against Lightning Discharges*, Chapters 1 to 5, www.itu.int/pub/T-HDB-EMC.3-1974-P1/en; Chapters 6 to 8, www.itu.int/pub/T-HDB-EMC.3-1978-P2/en; Chapters 9 and 10, www.itu.int/pub/T-HDB-EMC.3-1994-P3/ento

[557] Many EMC standards published by the European Telecommunication Standards Institution (ETSI), especially for wireless and/or radio communications, visit http://www.etsi.org/standards

Annex B.13   Some 'Ad Hoc' test methods

Note: *This is not an exhaustive list.*

[600] On-Site (in-situ) EMC Testing, Technical Guidance Note 49 from the EMC Test Laboratories Association, www.emctla.co.uk/technical-guidance-notes.aspx

[601] *Developing Immunity Testing to Cover Intermodulation*, W. Grommes and K. Armstrong, IEEE 2011 Int'l EMC Symp. Long Beach, CA, August 15-19, ISBN: 978-1-45770810-7

[602] *Testing for immunity to simultaneous disturbances and similar issues for risk managing EMC*, K. Armstrong, IEEE 2012 Int'l EMC Symp. Pittsburgh, PA, USA, August 5-10 2012, ISBN: 978-1-4673-2059-7

[603] *Using EMC HALT for risk and fault assessment*, Per Thaastrup Jensen, Proceedings of the 2013 International Symposium on Electromagnetic Compatibility (EMC Europe 2013), Brugge, Belgium, September 2-6, 2013, ISBN 978-1-4673-4980-2

[604] EMC Testing (in seven parts), *'Do-It-Yourself' testing from lowest-cost to fully accredited*, Keith Armstrong and Tim Williams, EMC & Compliance Journal, 2001-2002, from the 'Publications & Downloads' page at www.cherryclough.com

Annex B.14   Assessing the electromagnetic environment, and detecting threats

Note: *This is not an exhaustive list.*

[650] *Assessing an Electromagnetic Environment*, Technical Guidance Note 47 from the EMC Test Laboratories Association, www.emctla.co.uk/technical-guidance-notes.aspx

[651] Guides on 17 different electromagnetic phenomena and their EMC tests (including how to extend them to provide better 'coverage' of real-life electromagnetic disturbances), Keith Armstrong, REO (UK) Ltd., all free from www.reo.co.uk/technical_resources

[652] IEC/TR3 61000-2-1 *EMC: Description of the environment - Electromagnetic environment for low-frequency conducted disturbances and signalling in public power supply systems*

[653] IEC 61000-2-2 *EMC: Description of the environment- Compatibility levels for low-frequency conducted disturbances and signalling in public low-voltage power supply systems* – An IEC Basic EMC Publication

[654] IEC/TR3 61000-2-3 *EMC: Description of the environment - Radiated and non-network-frequency-related conducted phenomena* - An IEC Basic EMC Publication

[655] IEC 61000-2-4 *EMC: Description of the environment - Compatibility levels in industrial plants for low-frequency conducted disturbances* - An IEC Basic EMC Publication

[656] IEC/TR2 61000-2-5 *EMC: Description of the environment - Classification of electromagnetic environments* - An IEC Basic EMC publication

[657] IEC/TR3 61000-2-6 *EMC: Description of the environment - Assessment of the emission levels in the power supply of industrial plants as regards low frequency conducted disturbances*

[658] IEC/TR3 61000-2-7 *EMC: Description of the environment - Low frequency magnetic fields in various environments*

[659] IEC 61000-2-9 *EMC: Description of HEMP environment – Radiated disturbance*

[660] IEC 61000-2-10 *EMC: Description of HEMP environment – Conducted disturbance*

[661] IEC 61000-2-11 *EMC: Classification of HEMP environments*

[662] IEC 61000-2-13 *EMC: High power electromagnetic (HPEM) environments – Radiated and conducted*

[663] CIGRE 535, *EMC within Power Plants and Substations*, Working Group C4.208 April 2013 [Note: Section 6.3.1 Power system functions and corresponding acceptable degradation due to electromagnetic disturbances.]

[664] *A Cost-Efficient System for Detecting an Intentional Electromagnetic Interference (IEMI) Attack*, J. F. Dawson, I. D. Flintoft, P. Kortoci, L. Dawson, A.C. Marvin, M. P. Robinson, International Symposium on Electromagnetic Compatibility (EMC Europe 2014), Gothenburg, Sweden, September (2014)

[665] *Lessons Learnt From IEMI Detector Deployments*, D. Herke, L. Chatt, B. Petit and R. Hoad, 2016 European Electromagnetics (EUROEM) Symposium, London, July 11-14 2016

[666] *Autonomous Electromagnetic Attacks Detection considering a COTS Computer as a Multi-Sensor System*, C. Kasmi, J. Lopes-Esteves, M. Renard, General Assembly and Scientific Symposium (URSI GASS), 2014 XXXIth URSI, Page(s):1 – 4, 16-23 Aug. (2014)

[667] *EMC/EMI and Functional Safety: Methodology to characterize effects of interferences on devices*, C. Kasmi, J. Lopes-Esteves, K. Armstrong, Asia-Pacific EMC (APEMC) Symposium, Shenzhen, China, May 2016

Annex B.15 Verification/validation and other techniques (not specifically related to electromagnetic disturbances)

Note: *This is not an exhaustive list, and new standards are constantly being created.*

[700] IEC 60300-1 *Dependability management - Part 1: Dependability management systems*

[701] IEC 60300-2 *Dependability management - Part 2: Guidelines for dependability management*

[702] IEC 60300-3-1 *Dependability management - Part 3-1: Application guide – Analysis techniques for dependability – Guide on methodology*

[703] IEC 60300-3-2 *Dependability management - Part 3-2: Application guide – Collection of dependability data from the field*

[704] IEC 60300-3-3 *Dependability management - Part 3-3: Application guide – Life cycle costing*

[705] IEC 60300-3-4 *Dependability management - Part 3: Application guide – Section 4: Guide to the specification of dependability requirements*

[706]   IEC 60300-3-5 *Dependability management - Part 3-5: Application guide – Reliability test conditions and statistical test principles*

[707]   IEC 60300-3-7 *Dependability management - Part 3-7: Application guide – Reliability stress screening of electronic hardware*

[708]   IEC 60300-3-9 *Dependability management - Part 3: Application guide – Section 9: Risk analysis of technological systems*

[709]   IEC 60300-3-10 *Dependability management - Part 3-10: Application guide – Maintainability*

[710]   IEC 60300-3-11 *Dependability management - Part 3-11: Application guide – Reliability centred maintenance*

[711]   IEC 60300-3-12 *Dependability management - Part 3-12: Application guide – Integrated logistic support*

[712]   IEC 60300-3-14 *Dependability management - Part 3-14: Application guide – Maintenance and maintenance support*

[713]   IEC 60300-3-15 *Dependability management - Part 3-15: Guidance to engineering of system dependability*

[714]   IEC 60300-3-16 *Dependability management - Part 3-16: Application guide - Guideline for the specification of maintenance support services*

[715]   IEC 60410 *Sampling plans and procedures for inspection by attributes*

[716]   IEC 60605-2 *Equipment reliability testing - Part 2: Design of test cycles*

[717]   IEC 60605-3-1 *Equipment reliability testing. Part 3: Preferred test conditions. Indoor portable equipment - Low degree of simulation*

[718]   IEC 60605-3-2 *Equipment reliability testing. Part 3: Preferred test conditions. Equipment for stationary use in weatherprotected locations - High degree of simulation*

[719]   IEC 60605-3-3 *Equipment reliability testing - Part 3: Preferred test conditions - Section 3: Test cycle 3: Equipment for stationary use in partially weatherprotected locations - Low degree of simulation*

[720]   IEC 60605-3-4 *Equipment reliability testing - Part 3: Preferred test conditions - Section 4: Test cycle 4: Equipment for portable and non-stationary use - Low degree of simulation*

[721]   IEC 60605-3-5 *Equipment reliability testing - Part 3: Preferred test conditions - Section 5: Test cycle 5: Ground mobile equipment - Low degree of simulation*

[722]   IEC 60605-3-6 *Equipment reliability testing - Part 3: Preferred test conditions - Section 6: Test cycle 6: Outdoor transportable equipment - Low degree of simulation*

[723]   IEC 60605-4 *Equipment reliability testing - Part 4: Statistical procedures for exponential distribution - Point estimates, confidence intervals, prediction intervals and tolerance intervals*

[724]   IEC 60605-6 *Equipment reliability testing - Part 6: Tests for the validity of the constant failure rate or constant failure intensity assumptions*

[725]   IEC 60706-2, *Guide on maintainability of equipment, Part 2 - Section Five: Maintainability studies during the design phase*

[726]   IEC 60706-3 *Guide on maintainability of equipment, Part 3 - Sections Six and Seven, Verification and collection, analysis and presentation of data*

[727]   IEC 60706-5, *Guide on maintainability of equipment - Part 5: Section 4: Diagnostic testing*

[728]   IEC 61703 *Mathematical expressions for reliability, availability, maintainability and maintenance support terms*

[729]   IEC 62198 *Project risk management – Application guidelines Risk assessment methods*

[730]   IEC 62347 *Guidelines for establishing criteria for system dependability specifications*

[731]   IEC 62402 *Obsolescence management - Application guide*

[732]   IEC 61882 *Hazard and operability studies (HAZOP studies) - Application guide*

[733]   IEC 61078 *Analysis techniques for dependability - Reliability block diagram method*

[734]   IEC 61165 *Application of Markov techniques*

[735]   IEC 62308 *Reliability assessment methods*

[736]   IEC 60812 *Assessment techniques for system reliability – procedure for failure mode and effects assessment (FMEA)*

[737]   IEC 61025 *Fault tree assessment (FTA)*, also https://en.wikipedia.org/wiki/Fault_tree_analysis

[738]   Event Tree Analysis, https://en.wikipedia.org/wiki/Event_tree_analysis

[739]   IEC 61069-5 *Industrial-process measurement and control – Evaluation of system properties for the purpose of system assessment – Part 5: Assessment of system dependability (for the Fault Insertion Testing method)*

[740]   *The Cause Consequence Diagram Method as a Basis for Quantitative Accident Analysis*, B. S. Nielsen, Riso-M-1374, 1971.

[741]   Brainstorming techniques: https://en.wikipedia.org/wiki/Brainstorming; www.yourarticlelibrary.com/management/4-techniques-for-group-decision-making-process-more-effective/3506

[742]   SWIFT, Structured What-If technique, https://en.wikipedia.org/wiki/Structured_What_If_Technique

[743]   *Review of human reliability assessment methods*, Health and Safety Executive, 2009, http://www.hse.gov.uk/research/rrpdf/rr679.pdf

[744]   Task Analysis and Hierarchical Task Analysis (HTA), https://en.wikipedia.org/wiki/Task_analysis

[745]   Monte-Carlo methods, https://en.wikipedia.org/wiki/Monte_Carlo_method

[746]   Some references on Common Cause (sometimes called common-mode) failure analysis:

        https://en.wikipedia.org/wiki/Common_cause_and_special_cause_(statistics)#Common_mode_failure_in_engineering;

        Common-Mode Failure Considerations in High-Integrity C&I Systems, Thomson, Jim (February 2012) (PDF). Safety in Engineering;

        Randell, B. *Design Fault Tolerance*, in: *The Evolution of Fault-Tolerant Computing,(Dependable Computing and Fault-Tolerant Systems, Vol. 1)*, Avizienis, A.; Kopetz, H.; Laprie, J.C. (eds.), pp. 251-270. Springer-Verlag, 1987. ISBN 3-211-81941-X;

        SEI Framework: Fault Tolerance Mechanisms. Redundancy Management. NIST High Integrity Software Systems Assurance. March 30, 1995.

        A Study of Common-Mode Failures, Edwards, G. T.; Watson, I. A. (July 1979). SRD R146 (UK Atomic Energy Authority: Safety and Reliability Directorate).

        Defences against Common-Mode Failures in Redundancy Systems – A Guide for Management, Designers and Operators, Bourne, A. J.; Edwards, G. T.; Hunns, D. M.; Poulter, D. R.; Watson, I. A. (January 1981). SRD R196 (UK Atomic Energy Authority: Safety and Reliability Directorate).

[747]   *Time and Petri Nets*, https://en.wikipedia.org/wiki/Petri_net

        https://en.wikipedia.org/wiki/Coloured_Petri_net

        https://en.wikipedia.org/wiki/Stochastic_Petri_net

        https://en.wikipedia.org/wiki/TAPAAL_Model_Checker (for "timed-arc" Petri nets)

        Time and Timed Petri Nets, Serge Haddad, http://www.lsv.ens-cachan.fr/~haddad/disc11-part1.pdf

        Introduction to Petri Nets, http://neo.dmcs.p.lodz.pl/oom/petri_nets.pdf

[748]   IEC 61160, *Design review*

[749]   *Markov Models:*

https://en.wikipedia.org/wiki/Markov_model

https://en.wikipedia.org/wiki/Markov_chain

https://en.wikipedia.org/wiki/Markov_process

https://en.wikipedia.org/wiki/Layered_hidden_Markov_model

https://en.wikipedia.org/wiki/Maximum-entropy_Markov_model

https://en.wikipedia.org/wiki/Hierarchical_hidden_Markov_model

https://en.wikipedia.org/wiki/Hidden_semi-Markov_model

[750] *Reliability Block Diagrams*, https://en.wikipedia.org/wiki/Reliability_block_diagram

[751] *Cause consequence diagrams*, also known as Ishekawa or Fishbone diagrams

The Cause Consequence Diagram Method as a Basis for Quantitative Accident Analysis. B. S. Nielsen, Riso-M-1374, 1971

https://en.wikipedia.org/wiki/Ishikawa_diagram

[752] *Worst Case Analysis*, https://en.wikipedia.org/wiki/Worst-case_circuit_analysis

[753] *Preliminary Hazard Analysis*, Marvin Rausand, http://frigg.ivt.ntnu.no/ross/slides/pha.pdf

[753] *Hazard Identification: Review of Hazard Identification Techniques*, Health and Safety Executive, 2009, HSL/2005/58, http://www.hse.gov.uk/research/hsl_pdf/2005/hsl0558.pdf

[754] Taguchi's *Quality Engineering Handbook*, ISBN: 978-0471413349

[This is a very large book, about 1600 pages, but covers both concept and case study. It is truly a reference book, and would not be good for learning.]

[755] *Quality Engineering Using Robust Design*, Madhav S. Phadke, Prentice Hall, 1989, ISBN: 978-0137451678

ANNEX C

Some Functional Safety standards based on IEC 61508

IEC 61511	*Safety Instrumented Systems for the Process Industry Sector* (in USA: ANSI/ISA S84)
IEC 62061	*Safety of Machinery*
IEC 62278 / EN 50126	*Railways – Specification and Demonstration of Reliability, Availability, Maintainability and Safety (RAMS)*
IEC/EN 50128	*Software, Railway Control and Protection*
IEC/EN 50129	*Railway Signalling*
IEC 61513	*Nuclear Power Plant Control Systems*
RTCA DO-178C	*North American Avionics Software "Software Considerations in Airborne Systems and Equipment Certification"*
RTCA DO-254	*North American Avionics Hardware*
EUROCAE ED-12B	*European Flight Safety Systems*
ISO 26262	*Automobile Functional Safety*
ISO 26262-1	*Road vehicles -- Functional safety -- Part 1: Vocabulary*
ISO 26262-2	*Road vehicles -- Functional safety -- Part 2: Management of functional safety*
ISO 26262-3	*Road vehicles -- Functional safety -- Part 3: Concept phase*
ISO 26262-4	*Road vehicles -- Functional safety -- Part 4: Product development at the system level*
ISO 26262-5	*Road vehicles -- Functional safety -- Part 5: Product development at the hardware level*
ISO 26262-6	*Road vehicles -- Functional safety -- Part 6: Product development at the software level*
ISO 26262-7	*Road vehicles -- Functional safety -- Part 7: Production and operation*
ISO 26262-8	*Road vehicles -- Functional safety -- Part 8: Supporting processes*
ISO 26262-9	*Road vehicles -- Functional safety -- Part 9: Automotive Safety Integrity Level (ASIL)-oriented and safety-oriented analyses*
IEC 62304	*Medical Device Software*
IEC/EN 50402	*Fixed Gas Detection Systems*
DEF STAN 00-56	*Accident Consequence (UK military)*
IEC 60601-1-2	*Medical devices -- Application of risk management to medical devices,* (based on ISO 14971 instead of IEC 61508, but follows the same general risk management principles as IEC 61508)

ANNEX D

Glossary of terms, definitions and abbreviations

These descriptions are provided as an aid to understanding this Code of Practice.

Formal definitions for many of the terms may be found in the IEC International Electrotechnical Vocabulary.

AC	'Alternating current', a term used to denote electrical power or signals that are at a frequency other than 0 Hz.
CE marking	A form of mark that indicates that a product is claimed by its supplier to comply with all relevant EU Directives, such as the EMC Directive [31].
CISPR	Comité Internationale Speciale des Perturbations Radioélectriques, a branch of the IEC devoted to producing EMC test standards, usually (but not always) for emissions.
CM	'Common mode', a term used to describe voltages and/or currents that apply identically to all the conductors (including return conductors and shields) associated with a cable, or with an item of equipment, with respect to some remote reference. CM voltages or currents are always unwanted noise, and are associated with many EMC issues.
Competence	Having the training, technical knowledge, experience and qualifications relevant to the specific duties to be performed. (Adapted from IEC 61508-1, for more detail see IEC 61508-1 Ed.2:2010, sub-clauses 6.2.13 to 6.2.15)
Competent	Having the appropriate competence relevant to the specific duties to be performed. (Adapted from IEC 61508-1, for more detail see IEC 61508-1 Ed.2:2010, sub-clauses 6.2.13 to 6.2.15.)
Conducted	When applied to emissions or immunity, this term refers to unwanted electromagnetic energy conducted from equipment via the power supply or data, signal or control conductors.
Conducted emissions	Energy transmitted as electromagnetic waves along a cable or other conductor. Most countries have mandatory limitations on conducted emissions into their electrical power supply networks, to help reduce interference with other electronic equipment. Because conducted electromagnetic waves are a cause of radiated electromagnetic waves, these limitations also help to protect licensed users of the radio spectrum.
Conducted transients	Conducted emissions that are transient (short-term) in their nature, such as 'spikes', usually described in time-domain terms, for example, as a waveform, rather than frequency-domain terms, such as a spectrum.
Continuous disturbance	A disturbance that cannot be resolved into a succession of distinct events by measuring equipment. For transient disturbances, this term is typically applied to disturbances that occur more than 30 times a minute on average.
DC	'Direct current', a term used to denote an electrical power or signal voltage or current at 0 Hz.
Dip	A momentary reduction in the voltage of an AC or DC electrical power supply, usually for a time period of less than one second.

DM	'Differential mode', the mode of conduction of voltages and/or currents associated with intentional (wanted) power, signals, data, etc. A DM voltage is created on a conductor with respect to a different one in the same cable or item of equipment. A DM current generally flows through one conductor and returns by a different one in the same cable or item of equipment.
Dropout	A sudden reduction of the electrical power supply voltage to zero for a short period of time, usually less than 1 second, followed by a recovery to the original level.
DS	'Defined state', an equipment 'performance criterion': a specified, detectable operational state that an element or sub-system of a safety-related system switches to, temporarily or permanently within a stated time, if it suffers from errors, malfunctions or failures, for example, from EMI during an immunity test. Destruction of components is allowed as long as the DS of the EUT is maintained or achieved within a stated time. See 3.1.21 in IEC 61000-1-2:2016 [2].
	Some EMC publications related to functional safety used to use the abbreviation FS for this performance criterion, instead of DS.
Earth electrode	A conductor embedded in the soil beneath a site or building to try to make a low-impedance connection to the mass of the planet. Sometimes called a ground electrode.
Earthing	Sometimes called grounding; this is an electrical safety engineering term. Because RF references are often also developments of existing protective earthing structures, the action of connecting to an RF reference is often also called 'earthing'.
E/EE/PE	Electrical, electronic or programmable electronic (from IEC 61508).
EDR	Event data recorder, a non-volatile memory that stores 'events' detected by programmed diagnostic techniques, such as errors or failures, whether or not they were associated with a safety incident/accident.
Electromagnetic	All electrical and electronic phenomena (signals, data, power, etc.), radio waves, and light are electromagnetic in nature – their energy flows as a combination of both electric and magnetic energies.
Electromagnetic compatibility	'Electromagnetic compatibility', the ability of equipment to function satisfactorily in its electromagnetic environment without introducing intolerable electromagnetic disturbances to other equipment in that environment.
	(The EMC Directive, 2014/30/EU, Article 3.)
Electromagnetic disturbance	An electromagnetic phenomenon that, in a specified situation, could cause EMI.
Electromagnetic environment	The totality of the continuous and transient electric, magnetic and electromagnetic fields, conducted electromagnetic energies and electrostatic discharges at a given location at a given time.
Electromagnetic field	As an electromagnetic wave propagates in three-dimensional space and time, the magnitudes of its electric and magnetic waves can be represented as varying fields within the volume through which it is passing or has passed. Electric field strengths are measured in Volts/metre (V/m) and magnetic field strengths in Amps/metre (A/m).
Electromagnetic phenomenon	Any type of propagating electromagnetic energy (conducted, radiated, continuous, transient, electric, magnetic, voltage, current, common-mode, differential-mode, antenna-mode, arc or spark, etc.).
Electromagnetic wave	All electromagnetic energy travels in the form of waves, whether it is associated with electrical power, signals, data or control. In a conducted electromagnetic wave, the magnitudes of the voltages and currents vary along the conductor. In a radiated electromagnetic wave the magnitudes of the electric and magnetic fields vary with position in three-dimensional space.

Element	A part of a system comprising a single component or any group of components that performs one or more element safety functions.
	Note 1: An element may comprise hardware and/or software.
	Note 2: A typical element is a sensor, programmable controller or final element.
	(IEC 61000-1-2:2016 definition 3.1.12)
EMC	Electromagnetic compatibility.
EMC Directive	Legal instrument by which all member states in the European Union (EU) are obliged to enact national laws that have the same effect, to restrict the supply of electrical and electronic goods in the EU to those that meet certain minimum requirements for electromagnetic emissions and immunity. [31]
EMI	'Electromagnetic interference', sometimes simply 'interference'.
	The degradation in performance, malfunction or damage that is the result of inadequate immunity to electromagnetic disturbances.
EMP	'Electromagnetic pulse', a powerful radiated transient electromagnetic disturbance, sometimes used as shorthand for NEMP.
Equipment	A general term that refers to a wide variety of possible elements, modules, devices and assemblies of products (see IEC 61000-1-2:2016 definition 3.1.14).
ESD	'Electrostatic discharge', a sudden transfer of electric charge from one body to another, usually because of the voltage breakdown of the air between them (a spark). The dissipation of the charge causes transient disturbing currents to flow, and the spark is a source of very wideband radiated emissions.
ETSI	European Telecommunications Standards Institute, www.etsi.org.
EUC	'Equipment under control', equipment, machinery, apparatus or plant used for manufacturing, process, transportation, medical or other activities, including the EUC's control systems (see IEC 61508-4, definition 3.2.1, modified).
Far field	The far field of a source of electromagnetic radiation (i.e. non-ionising) is where the wave impedances of its propagating waves are constant and defined solely by the permeability and permittivity characteristics of the medium of propagation. Also see 'near field'.
Field	See 'electromagnetic field'.
Filter	A combination of capacitors, inductors, RF absorbers and/or resistors intended to reduce the amount of electromagnetic energy at certain frequencies from being conducted along a cable or wire.
FPGA	'Field-programmable gate array', an integrated circuit designed to be configured by a customer or a designer after manufacturing – hence 'field-programmable'. The FPGA configuration is generally specified using a hardware description language (HDL), similar to that used for an application-specific integrated circuit (ASIC). (Circuit diagrams were previously used to specify the configuration, as they were for ASICs, but this is increasingly rare.)
	FPGAs contain an array of programmable logic blocks, and a hierarchy of reconfigurable interconnects that allow the blocks to be 'wired together', like many logic gates that can be inter-wired in different configurations. Logic blocks can be configured to perform complex combinational functions, or merely simple logic gates like AND and XOR. In most FPGAs, logic blocks also include memory elements, which may be simple flip-flops or more complete blocks of memory. (From Wikipedia: https://en.wikipedia.org/wiki/Field-programmable_gate_array)

Functional safety	That part of the overall safety that depends on the correct functioning of electrical, electromechanical or electronic (hardware and software) technologies.
GHz	'Gigahertz', units of thousands of millions (10^9) cycles per second.
Grounding	Sometimes called earthing, this is an electrical safety engineering term. Because RF References are often also developments of existing protective grounding structures, the action of connecting to an RF Reference is often also called 'grounding'.
GSM	'Global system for mobile communications' (originally Groupe Spécial Mobile), the normal digital cellphone system, called GSM-850 and GSM-1900 in the USA, and GSM900, GSM1800 everywhere else, the numbers reflecting the frequency range of operation.
Harmonics	Frequencies that are integer multiples of the fundamental frequency. In AC mains electricity supplies they are caused by the power supplies of equipment drawing current in a non-sinusoidal manner, which distorts the waveform. All repetitive non-sinusoidal waveforms can be represented as the sum of a number of its harmonics, with various amplitudes and phases applied to each harmonic.
HEMP	'High-altitude electromagnetic pulse', electromagnetic pulse produced by a nuclear explosion outside the earth's atmosphere. Note: Typically above an altitude of 30 km. [see IEC61000-1-3:2002: *Electromagnetic compatibility (EMC): General - The effects of high-altitude EMP (HEMP) on civil equipment and systems*]
Hertz	Cycles per second, a measure of frequency.
HF	'High frequency', generally between 3 MHz and 30 MHz.
HR	'Highly recommended'
HV	'High voltage', in general usage: anything above 1 kV rms AC, or 1.5 kV peak DC. According to IEC standards: anything above 33 kV AC rms or 46 kV DC.
Hz	'Hertz'
IC	'Integrated circuit', a type of semiconductor device that contains many transistors, arranged to provide certain electronic functions. The latest types of IC can contain several million individual transistors.
IEC	'International Electrotechnical Commission', creates standards for EMC emissions and immunity, and safety, amongst many other issues, www.iec.ch.
IEMI	'Intentional electromagnetic interference', intentional malicious generation of electromagnetic energy introducing noise or signals into electric and electronic systems, thus disrupting, confusing or damaging those systems for terrorist or criminal purposes (taken from IEC 61000-2-13:2004 *Electromagnetic compatibility (EMC): Environment - High-power electromagnetic (HPEM) environments – radiated and conducted*).
IET	'Institution of Engineering and Technology', Savoy Place, London WC2R 0BL, created in 2006 by the merger of the IEE with the Institution of Incorporated Engineers (IIE), which dates back to 1884.
Interference	Usually shorthand for 'electromagnetic interference'.
I/O	Input/output.
ISM	A number of frequency bands set aside by international treaties for use by industry, science or medicine. There are no licensed radiocommunications in these bands, so the electromagnetic disturbances created by ISM equipment or systems should cause no interference with licensed users of the radio spectrum.

However, the levels of electromagnetic emissions permitted in the ISM bands by the relevant emissions standard (CISPR11) can be very high indeed, sufficient to cause health hazards to personnel, and to interfere with almost any kind of nearby electronic (possibly even electrical) devices, equipment and systems. Immunity to electromagnetic disturbances from nearby ISM equipment is not covered by any EMC standards in the IEC 61000-4 series, or listed under the EMC Directive.

Jamming	"A radio jammer is any device that deliberately blocks, jams or interferes with authorized wireless communications. In somecases jammers work by the transmission of radio signals that disrupt communications by decreasing the signal-to-noise ratio. The concept can be used in wireless data networks to disruptinformation flow. Jamming is usually distinguished from interference that can occur due to device malfunctions or other accidental circumstances. Devices that simply cause interference are regulated under different regulations. Unintentional 'jamming' occurs when an operator transmits on a busy frequency without first checking whether it is in use, or without being able to hear stations using the frequency.Another form of unintentional jamming occurs when equipment accidentally radiates a signal, such as a cable television plant that accidentally emits on an aircraft emergency frequency." Visit: https://en.wikipedia.org/wiki/Radio_jamming
JTAG	IEEE Standard 1149.1-1990, *Standard Test Access Port and Boundary-Scan Architecture.*
LEMP	'Lightning electromagnetic pulse', EMP caused by lightning strikes (either cloud-to-ground, or cloud-to-cloud). One of the means by which thunderstorms worldwide cost billions in damaged electronic equipment every year.
Lifecycle	A period of time that starts at the concept phase of a project and finishes when all of the safety-related systems and other risk reduction measures are no longer available for use (see IEC 61508-4, definition 3.7.1; excerpt modified).
Lightning protection	Protection against the direct and/or indirect effects of lightning.
Liveness	In concurrent computing, liveness refers to a set of properties of concurrent systems that require a system to make progress despite the fact that its concurrently executing components (processes) may have to 'take turns' in critical sections, parts of the program that cannot be simultaneously run by multiple processes. Liveness guarantees are important properties in operating systems and distributed systems. See [128].
MHz	'Megahertz', units of millions (10^6) of Hz.
Microsecond	10^{-6} seconds. One microsecond (μs) is one-millionth of a second.
Microwave	Typically, the frequency range above 1 GHz.
Millisecond	10^{-3} seconds. One millisecond (ms) is one-thousandth of a second.
Near field	The near field of a source of electromagnetic radiation (i.e. non-ionising) is where the wave impedances of its propagating waves are not stable with respect to the distance from their source, and the ratio of electric (E) field to magnetic (H) field can have various values that are not defined by the permeability and permittivity characteristics of the medium of propagation. Also see 'far field'.
NEMP	'Nuclear electromagnetic pulse', all types of EMP produced by a nuclear explosion [see IEC 61000-2-9:1996 *Electromagnetic compatibility (EMC: Environment: Description of HEMP environment — Radiated disturbance*]. Also see 'HEMP'.

Packet sequencing	"A powerful tool that uses the Sequenced Packet Protocol (SPP): a networking protocol that provides reliable transport of packets with flow control in environments where multiple transport connections are established. SPP uses destination ID reference numbers to identify the target end of a transport connection; sequence numbers to keep transmitted packets in the order in which they were sent; and acknowledge numbers that are assigned to the last packet in a sequence that a destination received properly to indicate that the transmission is complete and successful."

(From www.webopedia.com/TERM/S/sequenced_packet_protocol.html.) |
PCB	'Printed circuit board', laminated structure with layers of etched foil conductors (usually copper) known as tracks or traces, interspersed with layers of dielectrics (often a glass-fibre matrix). Sometimes called a printed wiring board (PWB). The traces are interconnected between layers by plated-through holes (PTH) known as 'via holes'. Electronic components are mounted onto the PCB and soldered to the traces on the outermost layer(s). Components with long pins or leads may be connected directly to traces on inner layers by plated through holes.
Power quality	A general term embracing a number of issues affecting the quality of the AC or DC electrical power supply, such as dips, dropouts, interruptions, sags, swells, harmonic waveform distortion, inter-harmonic waveform distortion, surges, spikes and transients. The standard for instruments measuring power quality is IEC 61000-4-30.
Radiated emissions	Energy transmitted as electromagnetic waves in the air or other dielectrics. Most countries have mandatory limitations on radiated emissions, to help protect licensed users of the radio spectrum from EMI.
Radiated transients	Radiated emissions that are transient (short-term) in their nature, such as 'spikes'. Usually described in time-domain terms, for example, as a waveform rather than frequency-domain terms (such as a spectrum).
RF	'Radio frequency', frequencies generally considered to be between 150 kHz and 300 GHz.
RF Reference	A conductive structure, usually a continuous or meshed (gridded) metal sheet or volume; in installations usually a meshed structure made of interconnected conductors and metal structures, that maintains a low impedance (generally much less than 1 Ω) up to some defined frequency.
Safety	Freedom from unacceptable risk (see ISO/IEC Guide 51:1999, definition 3.1).
Safety case	A structured argument, supported by a body of evidence, that provides a compelling, comprehensible and valid case that a safety-related system is safe enough for a given application in a given environment.
Safety documentation	Includes all the information necessary for the safe use of the item.

Safety manuals and safety cases are examples of such documentation. |
Safety integrity level (SIL)	A discrete level (one out of a possible four), corresponding to a range of safety integrity values, where safety integrity level 4 has the highest safety integrity level and safety integrity level 1 has the lowest. The target failure measures for the four safety integrity levels are specified in Tables 2 and 3 of IEC 61508-1 (see IEC 61508-4, definition 3.5.8).
Safety manual	For compliant items, a document that provides all the information relating to the functional safety of an element, in respect of specified element safety functions, that is required to ensure that the system meets the requirements of IEC 61508 series (see IEC 61508-4:2010, definition 3.8.17).
Safety-related system	The designated system that both:

– implements the required safety functions necessary to achieve or maintain a safe state for the EUC; and

– is intended to achieve, on its own or with other safety-related systems and other risk reduction measures, the necessary safety integrity for the required safety functions.

(See (IEC 61508-4, definition 3.4.1.)

SC	'Systematic capability', the measure (expressed on a scale of SC 1 to SC 4) of the confidence that the systematic safety integrity of an element meets the requirements of the specified SIL, in respect of the specified element safety function, when the element is applied in accordance with the instructions specified in the compliant item safety manual for the element.

Note 1: Systematic capability is determined with reference to the requirements for the avoidance and control of systematic faults (see IEC 61508-2 and IEC 61508-3).

Note 2: What is a relevant systematic failure mechanism will depend on the nature of the element. For example, for an element comprising solely software, only software failure mechanisms will need to be considered. For an element comprising hardware and software, it will be necessary to consider both systematic hardware and software failure mechanisms.

Note 3: A Systematic capability of SC N for an element, in respect of the specified element safety function, means that the systematic safety integrity of SIL N has been met when the element is applied in accordance with the instructions specified in the compliant item safety manual for the element.

Screening	An alternative term for shielding.
Shielding	The use of conducting material to form a barrier to electromagnetic waves so that they are reflected and/or absorbed. Also known as screening.
Surge	A type of transient voltage band/or current with a high energy content, typically produced by the energy associated with a lightning strike or flyback of stored inductive energy (for example, in a large electric motor or generator) coupling into cables such as power supply or telecommunication cables. A surge is generally considered to have much longer rise times and decay times, and have much more energy associated with it, than a 'transient', 'fast transient' or 'spike'.
Surge protection device	A device for suppressing surges, typically by switching to a low-resistance state to shunt surge energy away from a protected circuit, such as an MOV, spark gap, TVS, SAD, etc. Sometimes called a surge arrester.
Transient	A rapid change of the waveshape of voltage, current, or field, of very short duration followed by a return to steady state. Usually described in time-domain terms, for example, as a waveform (rather than frequency-domain terms, for example, as a spectrum).
V/m	Volts/metre, the standard unit of electric field (E-field) strength.
Volt	The standard unit of measuring electrical voltage (potential difference).

ANNEX E

Comparisons with IEC 61508-7

This table shows the relationship between the techniques and measures listed in this Code of Practice with those in IEC 61508-7 Edition 2:2010. Additional techniques or measures that do not occur in IEC 61508-7 Ed.2:2010 are shown by an X in the right-hand column.

Section number in this Code of Practice	Equivalent, in IEC 61508-7 Ed.2:2010 Annex A	Equivalent, in IEC 61508-7 Ed.2:2010 Annex B	Equivalent, in IEC 61508-7 Ed.2:2010 Annex C	No equivalent in IEC 61508-7 Ed.2:2010
2.1.1		B.1.1, B.1.2		
2.1.2		B.2	C.2	
2.1.3		B.3.1		
2.1.4		B.3.1		
2.2.1		B.1.3		
2.2.2			C.5.2	
2.2.3	A.11.4	B.1.4	C.3.1, C.3.4, C.3.5	
2.2.4		B.5		
2.2.5				X
2.2.6.1	A.6.2		C.3.2	
2.2.6.2			C.3.2	
2.2.6.3			C.2.5	
2.2.7				X
2.2.8				X
2.2.9				X
2.2.10				X
2.2.11				X
2.3.1		B.4.1, B.4.2		
2.3.2		B.4.3		
2.3.3		B.4.4		
2.3.4		B.4.6		
2.3.5		B.4.8		
2.3.6.1			C.2.5	
2.3.6.2			C.2.5	
2.3.6.3			C.2.5	
2.3.7			C.2.6.5	

Section number in this Code of Practice	Equivalent, in IEC 61508-7 Ed.2:2010 Annex A	Equivalent, in IEC 61508-7 Ed.2:2010 Annex B	Equivalent, in IEC 61508-7 Ed.2:2010 Annex C	No equivalent in IEC 61508-7 Ed.2:2010
2.3.8			C.2.6.6	
2.3.9			C.2.6.7	
2.3.10.1	A.4.2, A.4.3, A.4.4			
2.3.10.2	A.4.5			
2.3.10.3			C.5.4	
2.3.11	A.2.1, A.2.5, A.4.5, A.5.7, A.6.3, A.7.3, A.11.4			
2.3.12	A.7.5			
2.3.13.1	A.5.1, A.5.2, A.5.3, A.5.4			
2.3.13.2	A.5.5			
2.3.13.3	A.5.7			
2.3.13.4			C.5.4	
2.3.14	A.4.1, A.5.6, A.6.2, A.7.1, A.7.2		C.3.2	
2.3.15.1	A.3.3			
2.3.15.2	A.3.4			
2.3.15.3	A.3.5			
2.3.15.4	A.3.1, A.3.2			
2.3.16	A.1, A.2			
2.3.17				X
2.3.18.1	A.2.1			
2.3.18.2	A.2.2			
2.3.18.3	A.2.3			
2.3.18.4	A.2.5		C.3.4	
2.3.18.5	A.2.6			
2.3.18.6	A.2.7			
2.3.18.7			C.2.5	
2.3.19.1	A.9.1			
2.3.19.2	A.9.2			
2.3.19.3	A.9.3			
2.3.19.4	A.9.4			
2.3.20	A.6.3, A.6.4, A.6.5			
2.3.21	A.6.1, A.7.4			
2.3.22				X
2.3.23.1	A.8.1, A.8.3			
2.3.23.2				X

Section number in this Code of Practice	Equivalent, in IEC 61508-7 Ed.2:2010 Annex A	Equivalent, in IEC 61508-7 Ed.2:2010 Annex B	Equivalent, in IEC 61508-7 Ed.2:2010 Annex C	No equivalent in IEC 61508-7 Ed.2:2010
2.3.23.3				X
2.3.23.4	A.8.2			
2.3.24	A.10			
2.3.25				X
2.3.26	A.11.1, A.11.2, A.11.3			
2.3.27		B.3.1		
2.3.28	A.2.8			
2.4.1		B.4.1		
2.4.2		B.4.1		
2.4.3		B.4.1		
2.4.4		B.4.1		
2.5.1		B.6	C.5, C.6	
2.5.2		B.5.1, B.5.2, B.6.1, B.6.2		
2.5.3		B.5.1, B.5.2, B.6.1, B.6.2, B.6.8		
2.5.4		B.4.1		
2.6.1		B.1.1		
2.6.2		B.1.1		
2.6.3		B.1.1		
2.7				X
2.8.1				X
2.8.2				X
2.8.3				X

INDEX

E

O

P

Q

R

S

IET Standards

Influence the future of Standards

Working in a fast paced, rapidly changing industry has its frustrations. A lack of professional standards and guidance increases risk, and hinders the ability to embrace innovation.

The IET uses its wealth of knowledge and experience to bring about standards that:

- Solve common working problems
- Make meeting legislative requirements simple
- Give practical guidance for practising engineers

To create the best possible guidance, we need you.

Get involved as a member of a publication committee, take on authorship of a book or simply give us feedback on a draft publication. The choice is yours.

Find out more at:

www.theiet.org/setting-standards

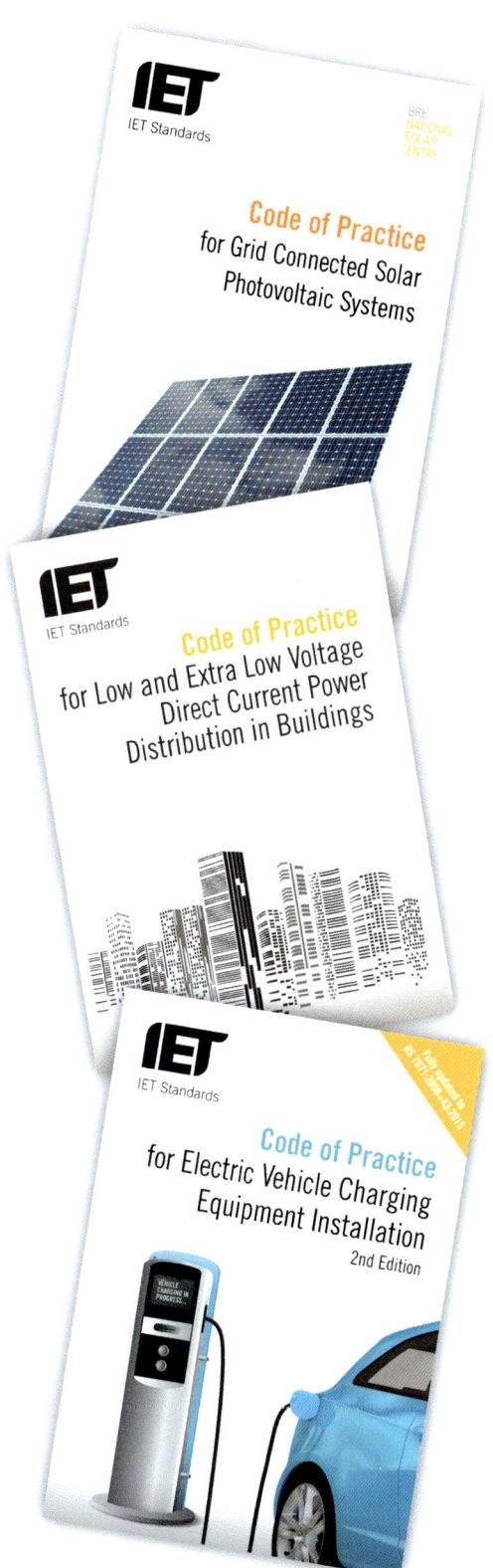

Electrical **excellence**

Expert publications

The IET is co-publisher of BS 7671 (IET Wiring Regulations), the national standard to which all electrical installations should conform. The IET also publishes a range of expert guidance supporting the Wiring Regulations.

You can view our entire range of titles including...

- BS 7671
- Guides
- Guidance Notes series
- Inspection, Testing and Maintenance titles
- City & Guilds textbooks and exam guides

...and more at:

www.theiet.org/electrical

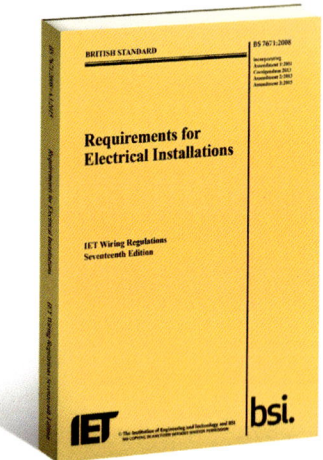

ELECTRICAL STANDARDS +

Constantly up-to-date digital subscriptions

Our expert content is also available through a digital subscription to the IET's Electrical Standards Plus platform. A subscription always provides the newest content, giving peace of mind that you are always working to the latest guidance.

It also lets you spread the cost of updating all your books once new versions are released.

Going digital gives you greater flexibility when working with the Wiring Regulations, Guidance Notes and the IET's expert Codes of Practice available for electrical engineers. The intuitive search function instantly serves results from across all books in your package. You can also access the content on your desktop, laptop or tablet, making it easy to take the content out on site or read on the move.

Find out more about our subscription packages and choose one to suit you at:

www.theiet.org/esplus